THE CATTLEYAS AND THEIR RELATIVES

Volume VI. The South American *Encyclia* Species

"Man may be defined to be a Cooking Animal," a quotation from James Boswell, matches well what James Bateman wrote in *The Orchidaceae of Mexico and Guatemala* (1843) about collecting and the habitats of orchids: "What a rich and redundant vegetation! . . . The happy pair in the foreground . . . are not confining themselves to a vegetable diet. That *Epidendrum* on the trunk of the prostrate tree must be a fine thing, and we are only sorry to see the lady turn her back upon it." Note also the monkey, snake, "parroquet," and the "cayman's awful head" among the "tangled luxuriance and dismal grandeur of the forest scenery of tropical America."

THE CATTLEYAS AND THEIR RELATIVES

Volume VI. The South American *Encyclia* Species

A book in six parts
by

CARL L. WITHNER

TIMBER PRESS
Portland, Oregon

ISBN 0-88192-436-9

Printed in Hong Kong

Timber Press, Inc.
The Haseltine Building
133 S.W. Second Avenue, Suite 450
Portland, Oregon 97204, U.S.A.

Library of Congress Cataloging-in-Publication Data
(Revised for vol. 6)

Withner, Carl L. (Carl Leslie)
The Cattleyas and their relatives.

Includes bibliographical references.
Includes indexes.
Contents: v. 1. The Cattleyas — v. 2. The Laelias — v. 3. Schomburgkia, Sophronitis, and other South American genera. — v. 4. The Bahamian and Caribbean Species. — v. 5. Brassavola, Encyclia, and other genera of México and Central America. — v. 6. The South American Encyclia Species.
1. Orchids. 2. Cattleyas. 3. Orchid culture. 4. Orchids—Latin America. I. Title.
SB409.W793 1988 635.9'3415 88-8560
ISBN 0-88192-099-1 (v. 1)
ISBN 0-88192-161-0 (v. 2)
ISBN 0-88192-269-2 (v. 3)
ISBN 0-88192-344-3 (v. 4)
ISBN 0-88192-456-3 (v. 5)
ISBN 0-88192-436-9 (v. 6)

Table of Contents

This volume is dedicated to
John Lindley (1799–1865)
and
Oakes Ames (1874–1950),
giants in the orchid world
without whom
this work could not have been achieved

Preface

Time has gone quickly since I started writing this series entitled *The Cattleyas and Their Relatives,* and I am anxious to get it finished as I approach my 82nd year. Not only are there other things to do, but I am getting to the stage where travel is more difficult and things seem to take twice as long to accomplish as they did formerly, and perhaps even longer. The subject matter is still fascinating to me, and I try to write these descriptions accurately, depending upon my large notebooks of data about plants I have seen in the wild, in shows, and in collections.

One finds, however, that regardless of the data at hand, a few details are always missing, and most often the herbarium record is minimal on many of these species. Frequently, especially with the genus *Encyclia* in South America, the original or type specimen may have been destroyed. We have been most fortunate that records on many of these types have been available at Harvard University through the early research of Oakes and Blanche Ames in various European herbaria and conserved at the Oakes Ames Orchid Herbarium. They photographed and made drawings of many type records that no longer exist in the original. The orchid herbarium at Kew is different, but equally important for this research, and the Lindley Herbarium there with its many early specimens, most of which are types, cannot be undervalued. The Reichenbach records in Vienna provide the third leg for the research that has been necessary for these books, and particularly for *Encyclia,* with which this volume is especially concerned.

I have put the bulk of the *Encyclia* complications off until Volume VI, but, finally, I have had to come face-to-face with the many problems and difficulties that this genus (as it occurs in South America) provides. No one has ever tried putting this information together for that continent. I have no illusions now why that is so, and I have no doubt that some of my material will be revised in the future. But, at this time, and with the data I have systematically accumulated in three volumes of herbarium notes and innumerable photocopies since the early 1960s, I can hope that the basic information will be available in one place, however it may eventually be interpreted.

In writing these pages I have tried to include explanations of why certain opinions about identifications, synonymies, or distributions come to hand. Certain technicalities of terminology are required, but the words seem to be necessary in presenting the material involved. Where there are deficiencies I have tried to note them so that future fieldworkers can write better descriptions and collect more plants before it is too late, and so that orchid floras for many countries can be completed in better detail.

Conservation of habitats comes constantly to the fore, along with the lack today of field botanists who know about orchids and the role of particular insect or other pollinators that are involved with these species in the wild. All these details represent gaps in our knowledge. I await the day when we can distinguish the plant species by running them through DNA analysis or some comparable biochemical scheme that will tell all, but I am not yet convinced that such techniques will answer all the questions. We must still deal with the living plants as they occur in nature.

At least the culture of most of these *Encyclia* species in man-made surroundings is simple enough so that living herbaria can be provided for future research. All one has to do is get the plants in the first place and before it is too late! These days many of our conservation laws and regulations have just the opposite effect and offer little remedy for dealing with local conservation difficulties and problems on an on-going basis. Enough said!

When I was a new student of botany, species were looked upon by most as created entities that had perpetuated themselves here on earth, man included, and they were wrapped around with moral and spiritual parameters. They were referred to later as Linnean or taxonomic species. After concepts of evolution and the beginnings of genetics came along, more dynamic definitions were necessary. Taxonomy, as someone has said, is comparatively easy when you only have one example to study! But, as populations instead of individuals began to be studied, and things like variation among individuals, color forms, mutations, extinctions, introgression with other species, hybridity both natural and by man, distributions in relation to habitats and geology, particularly disjunctions, and many other factors came to the fore—especially with plant species—the old species concept had to be modified.

The "biological species concept" gradually evolved, and we are still trying to fit the orchids into this open-ended idea. Botanists are now using biochemical observations to help explain the variations among members of similar populations; and it may be hoped that eventually DNA studies might help us further along in understanding the complex relationships among individuals. Species can no longer be placed in neat little boxes, signed, sealed, and delivered, especially, it would seem, orchid species that we have caught in the midst of what has been termed "explosive" evolution. The cattleyas, the laelias, the brassavolas, and even the encyclias, from these six volumes, would be fine subjects for biochemical species understandings. The hobby grower, to counterbalance the scientist, merely wants a name to put on a label that will not have to be changed in his or her lifetime. It will not happen soon, nor will the "lumpers" and "splitters" ever manage to view the mounting evidence about a given species or genus and come to the same conclusions. As one botanist has asked, "What do you get with three taxonomists looking at an orchid?" The answer, of course, is "Four opinions!"

As in previous volumes, I have numbers of the same people to thank for help they have provided over the years: Leslie Garay, Phillip Cribb, Eric Christenson, Antonio Toscano de Brito, Lou Christian Menezes, Denis Duveen, Jorge and Mauricio Verboonen, Sandra Odebrecht Champlin, Red and Trudi Marsh, Lillian Severin, Stalky and Nora Dunsterville, Ron Parsons, and Jack Fowlie. Eric Christenson has been very patient with my inquiries on technical nomenclatural problems, especially the matter of priorities with homonyms that have been a particular plague for this volume. A comparative newcomer to the appreciation list is David Hunt of Houston, Texas. He has done much collecting in faraway places, and grows and photographs many encyclias that I have not previously been able to see in living condition. I am particularly indebted for some of his photographs in this volume.

I am also indebted to the Orchid Digest Corporation for their loan of two of Dr. Fowlie's research notebooks on Brazilian *Encyclia* species. Many of his slides, taken by Trudi Marsh or by Lou Christian Menezes, have been of particular help, and some of them are reproduced here, courtesy of the *Orchid Digest* co-editors, Don Herman and Ernest Hetherington. The volunteer aid of Betty Dunton, on behalf of the *Orchid Digest,* has been cheerfully given in going through many of the Fowlie letter files on the encyclias. Much of the early work by Fowlie and Duveen was done through the facilities and varied collection of encyclias and other orchid species provided by the Verboonens at Orquidário Binot. Lou Menezes has also helped me directly by providing first-hand information as well as some of her slides. Through her orchid collection and contacts with various orchid societies and their enthusiasts in Brazil she has been a close collaborator on the Brazilian species. Without her photography and the records of the *Orchid Digest,* this volume would be sadly missing many of its color plates. The color photo acknowledgments will all be found in the plate captions. Just count them! Menezes, especially, has been responsible for a whole rash of new Brazilian encyclias, many more than anyone ever anticipated.

As this volume draws to a close, I must testify again to the help I have had from my wife, Pat. Not only has she tolerated all my rantings and ravings about how I ever became involved in writing such a series, she has suffered through my learning how to use a computer for word processing and databasing. Her early newspaper training as a copywriter, proofreader, and all around English major and wordsmith has provided untold help with everything from dangling modifiers to one of her favorites—misplaced apostrophes. I will say that she has not minded the field trips and library-herbarium travels that these books have entailed. Pat has collected orchid friends as I have collected data or plants.

Though many of the line illustrations have been culled and adapted from my research notes of many years, Michael Ashe has drawn the plates for the comparative studies of the *Encyclia* species in this volume. Those drawings, along with the comparable plates on *Encyclia* from Volumes IV and V, provide at a glance a survey of the genus—as I have discussed it earlier in Volume IV.

Of the 146 *Encyclia* species presented in these six volumes, 27 are from the Bahamas and the Caribbean, 41 from Mesoamerica, and with this volume about 78 more from South America. I already know of at least two or three new species, definitely more, which are not yet described. As they become published in orchid journals here and there, adding them to the lists in this series will be your homework assignment for the future.

At this point I know a little better what unconscious impulses motivated me to take pictures of so many flowers of the species of the Laeliinae at shows and in collections over the world. Seldom are sufficient drawings or photos available from which to make a selection for publication, and some slides have not been good at all—but they are better than nothing, the obvious alternative. Getting any photographs, especially publishable ones, has been the hardest part of producing these volumes. I have said it before: we need to set up a clearinghouse on the Internet for orchid photos, especially those of species that have been properly identified. I hope that this series may have provided a service in that regard, but with a lack of several photos in this volume, there are proportionately more black-and-white figures to help illustrate the species. The value and importance of type specimens and original sketches conserved in various herbaria cannot be overstated.

Once again I must commend the data provided with much patience and accuracy by Rudolf Jenny of Moosweg, Switzerland. He has been most generous and prompt in answering my cries for library help with his superior cross-indexed database of orchid literature. It has helped me connect with desired references whether I knew only people names, or dates, or the name of the plant. His references have been especially useful for this volume in the case of homonymic or homophonic names. I sometimes feel as though he should be added as a co-author, as his help with literature citations has been unending. Eric Christenson would have to be another co-author for checking my notations on nomenclature against his encyclopedic knowledge of the International Code and the validity of one name versus another. My third co-author would have to be Antonio Toscano de Brito from Brazil who made many of the type records into useful drawings for the details of flower anatomy. No single person could do this job alone—though some individuals have tried. In Europe it is referred to as the Reichenbach Syndrome.

I would also like to thank the people who have produced checklists for the orchids in various countries with which we are concerned in this volume. Particularly helpful to me were two lists for Colombia by Calaway Dodson, one in the book by Mariano Ospina H. (1996) and another in the book by Pedro Ortíz V. (1976), totaling 2982 species in 217 genera. Dodson's (1988) list for Ecuador contains 2425 species in 203 genera, plus 196 unidentified collections that could represent undescribed species or might be included in the 317 species reported for Ecuador but not found by Dodson and his co-workers, and, finally, 134 new species not yet included in Dodson's count. The race against time is on! The list for Venezuela by Ernesto Foldats (1970) and another for northeastern Argentina by Louis O. Williams (1939) must also be mentioned. The preliminary list by F. C. Hoehne (1952) of *Encyclia* species centered in Brazil has been most useful, as have the list by Porto and Brade (1935) of Brazilian species and another by Rodrigues da Silveira (1935). Finally, I was privileged to have seen the draft list, yet unpublished, by Eric Christenson and J. K. Boggan for the Guianas. Of course, at this point, I was concentrating only on the *Encyclia* species that were listed for South America, and they provide a good counterpoint to my *Encyclia* list that you will find here in Chapter 2.

Sometimes the herbarium record and various reports are not sufficient to disclose the actual distribution of a given taxon. The record, remember, ultimately depends upon correctly identified herbarium specimens from known localities. Some species may not be quite as endemic as they seem; and it is obvious that others have a more generalized distribution. Where endemic stops and a general distribution begins is not generally stated—if the difference is noted at all—but most of the species do inhabit definite areas that can probably be correlated, in turn, with the ancient geological history of the various regions.

Unfortunately, mistakes in spelling and book and journal references seem to show up clearly only after proofread manuscripts are published. Those that have turned up in these volumes are my responsibility. Fortunately they have diminished in number as the volumes have progressed, thanks mostly to the good help of Timber Press staff, particularly Linda Willms and Darcel Warren. I have been marking errata diligently in bright red ink as they are reported or otherwise turn up in my desk copies of the volumes. Some books have been reprinted without corrections as revised editions of these volumes would not be all that different in content to warrant the expense of new ones.

For now, Appendix 1 of Volume VI itemizes the nomenclatural details that were new among the species covered—an aid for the technical details that must be recorded for other taxonomists or botanists. Appendix 2 gives me an opportunity to deal with details of the descriptions that some have questioned, outright mistakes in citations, and points about which others may have provided new information. Both appendices are probably required reading for anyone using this set of books.

This sixth volume has particularly presented many difficult decisions about synonymies and the correct names for the species. Many type descriptions from years before do not contain pertinent data we should have today to decide—and type specimens are often not readily available for handy consultation. It is therefore not surprising that orchidologists may come to different conclusions and interpretations of the material that can be found. I have tried, at least, to have a drawing or a color plate for as many species as possible. If I am lucky I may have both, especially if the herbarium record includes the sketch of a flower dissection on the type sheet. For a few species I have neither.

Often there may be only a single herbarium sheet or two anywhere in the world upon which to base an opinion, and for some species there have never been drawings, photos, or illustrations of any kind. Furthermore, some specimens may have limited access at best. At worst, some to many of the specimens have been misidentified. So, be tolerant and understanding of what may be the best opinion possible at this time. If you have a rare or questionable *Encyclia* species to identify, photograph it carefully and press an inflorescence to send for conservation to a well-known herbarium, especially if you know about the plant's geographic origins. Gleanings from orchid shows could well be a source of rare or unusual items, particularly when they are in tropical countries or other areas where conservation or rarity is a particular consideration. I have often been frustrated by these *Encyclia* specimens and have transferred some of them back and forth among various species trying to find their proper niche. Some, no doubt, will end up as yet-to-be-described species; others will turn out to be old friends; still others may end up as synonyms—not a fate worse than death! The cumulative record is what counts.

I would sign off now with only one caveat—read and see with a probing and observing open mind! Just because all this material is written down in so many words in so many volumes do not think the subject matter is closed or that it can all be accepted exactly as written. Allow for new information and new opinions and interpretations as time provides new details. Let your opinions be modified as may be necessary, and do not practice what is known as "patriotic taxonomy!" Do not worry about the splitters and lumpers, but do write your flower pot labels as accurately as you can, including the country of origin whenever the source is known for your plant. The text for these six volumes is not yet finished.

January, 2000

CHAPTER 1

Regional Keys to the Genus *Encyclia*

Although I have explained *Encyclia* flower structure in Volumes IV and V, some points should be kept in mind when identifying an unknown species. When there are so many species in a given genus, it may be difficult to find a single hallmark for identification of each one, and usually it is a combination of details that shows the discontinuity between one species and the next. Keys for identification are made up of such details and are therefore a challenge to write or compose, and then a challenge to use. The same details must be at hand for any identification procedure.

One of the first things that has to be determined is if the species is actually in the key or the book. Because of their endemic nature, many encyclias can sometimes only be found in a key specific to a certain country or region. That means geography is always important for plant identification, especially for orchids. If you know where the plant came from, its country of origin, you are ahead of the game and can eliminate many species without further investigation. Otherwise, you must gradually eliminate one area after another, mostly by running the species through the various alternatives in the key. If the plant seems to correspond with the various characteristics in the key, keep on to the end—the name of the species. If you find that the key characters do not seem to fit the plant so that the plant does not key readily or without confusion, then consider that you may have ignored a proper choice early on in the key, or the plant may not be in the key. You may then have to start over, try another key or reference book, and re-key the plant until a proper name seems correct.

Before starting to use a key, look carefully at the plant and its flowers. What details must you observe before you start? Here is a list.

1. **Pseudobulbs:** Are they clustered or separated by rhizomatous extensions? What shape are they? What is their size? What color are they? Do they have any unusual feature?

2. Leaves: How many are on an average growth? What are their length and width? What shape are they? What shape is the leaf tip? Are the lower and upper leaf surfaces similar or different? Do the leaves have any outstanding quality, such as blunt tip with unequal halves, grooves, folds, leaf texture, position, or other feature?

3. Inflorescence: Is it an unbranched stalk (a raceme) or a branched flower cluster (a panicle)? Are the flowers widely spaced or crowded together on it? Are the flowers clustered at the tip or evenly spaced along the inflorescence? Is it erect, nodding, or hanging? Is its height below the leaf height or above?

4. Sepals and petals: Are they the same or different in shape, coloration, markings, and presentation? The dorsal sepal is the sepal most used for comparison with the petals.

5. Lip: Pull down the lip and remove it to flatten for examination, noting as you do what relationship the lateral lobes may have with the column. Flatten the lip by taping it down to a card with clear cellophane tape. Note the natural angle of the lateral lobes to the axis of the flattened lip. Some prefer using double-sided tape, spreading the lip naturally upon it. What are the overall length and width of the lip and its various parts? What is its spread (the distance across the lip from the tip of one lateral lobe to the other)? Is the base of the lip clawed? What are the shape, stance, and color of the lateral lobes, and what shape are its apices? Is the callus configuration typically forcipate or some variation thereof? Is there an unusual hirsute callus area? What veining pattern extends from the callus over the disc of the midlobe? What are the shape, coloration, and margin of the midlobe, and what shape is its apex? Is there an apiculus, a small point, at the tip of the main central vein?

6. Column: What is its configuration—length, straight, humped, S-shaped, curved upwardly—and angle with the ovary? Does it have auricles, horns, wings or extensions into teeth or auricle-like infoldings? What is the color of the column? What are the configuration and color of the anther cap?

7. Ovary: How long is the ovary and its stalk (the pedicellate ovary)? Is the surface smooth or somewhat roughened to the touch? What are the apex and base like?

8. Flower stalk: Is it roughened or smooth, straight or fractiflexed? What color is it? What position is it? How long is it?

9. Flowers: Are they fragrant? If so, is the fragrance noticeable in the day, or night, or both? What is the familiar or generic identification?

I have attempted to include such information itemized above in the descriptions of the species below. Sometimes the information is not mentioned in the type description, and if the specimen has been destroyed, or is not locatable or available, that information may be lacking.

By the way, most encyclia blooms seem to be midsized, though plants with smaller and larger flowers turn up. Larger-flowered types include *Encyclia alata, Enc. cordigera,* and *Enc. randiana.* Medium-flowered types include *Enc. diurna, Enc. oncidioides,* and *Enc. tampensis.* Small-flowered types include *Enc. microbulbon, Enc. microxanthina,* and *Enc. xipheres,* the sort with fewest examples in most herbaria. The distinction is arbitrary and for these purposes a small lip has a maximum length of 14–15 mm, with most from 10–12 mm; a large lip has a minimum length of 20–22 mm and may reach 50 mm. In between accounts for most species, the midsized forms.

Table I represents my attempt to sort the South American species by regions, although we do not know exactly the distribution of all the species concerned. Future research may enable us to correct or amplify this information, but this is as it stands at present. The listings include synonym names that should lead you to the currently accepted name as recorded in this volume. It is conventional for orchidologists concerned with classification problems and herbaria to sort specimens from South America into three major regions: the Andean and western areas; the eastern, predominantly Brazilian region; and sometimes the more northern region, including the Guianas, Venezuela, and Colombia. Naturally, there is overlap among the areas so it does not work perfectly all the time. I have subdivided the system in Table I to suit my purposes, and if there is overlap, the species may be entered more than once.

TABLE I

South American *Encyclia* Species and Their Synonyms Listed by Major Country(ies) of Distribution

The distributions are based, for the most part, upon herbarium specimens, but the list is subject to revision as identifications are checked and we learn more from fieldwork with these species. A question mark after the name indicates questionable or possible locations that may not be confirmed by actual conserved specimens. At this point, this list of species and synonyms is based on a variety of sources, including collectors' reports and collections by hobbyists that may not be confirmed by properly identified herbarium material. The locations of some species may turn out to be hypothetical until more collections are made, identified, and conserved in a herbarium. Although Table I may list species for a given area without a voucher specimen to substantiate it, the countries listed in the text with the species descriptions are mainly based on actual specimens from herbaria and type descriptions.

Argentina, Bolivia, and Paraguay

amicta
angustiloba
argentinensis
buchtienii = *steinbachii*
conchaechila
cyperifolia
ensiformis = *cyperifolia*
guentherianum = *angustiloba*
linearfolioides = *amicta*
mapiriensis = *angustiloba*
microtos var. *grandiflora* = *cyperifolia*
microxanthina?
pflanzii = *thrombodes*
saltensis = *argentinensis*
sclerocladia?
steinbachii
thrombodes
yauaperyensis

Ecuador and Peru

amicta
angustiloba
aspera
asperirachis = *aspera*
bacigalupianum = *maderoi?*
buchtienii?
chlorantha = *chloroleuca*
chloroleuca
cyperifolia
ensiformis = *cyperifolia*
guentherianum = *angustiloba*
maderoi?
mapiriensis = *angustiloba*
microtos
microtos var. grandiflora = *cyperifolia*
naranjapatensis
oncidioides var. *gravida* = *oncidioides*
pflanzii = *thrombodes*
pilosa
randiana?
replicata? = *selligera* (see Vol. V)
sclerocladia
stellata?
thienii = *chloroleuca*
thrombodes
trachypus = *microtos?*
xipheres?
yauaperyensis

Colombia

aspera
asperirachis = *aspera*
bacigalupianum = *maderoi*
ceratistes
chloroleuca
cordigera
diurna
expansa
garzonensis
insidiosa?
ionosma?
laxa?
leucantha
maderoi
magdalenae
microtos
oncidioides var. *profusa* = *profusa*
pflanzii = *thrombodes*
picta?
pilosa
profusa
replicata = *selligera* (see Vol. V)

TABLE I *(Continued)*

recurvata?
stellata
thienii = chloroleuca
thrombodes
trachypus = aspera

Venezuela

acuta
affinis = guianensis
amicta
amicta subsp. *conchaechila = conchaechila*
aspera?
asperirachis = aspera?
auyantepuiensis
bacigalupianum = maderoi
bradfordii
ceratistes
chloroleuca
conchaechila
cordigera
diota = granitica?
diurna
ensiformis? = oncidioides
flava?
glandulosa = diurna
granitica
guianensis
halatum = recurvata
hermentiana
ivonae
latipetala
leucantha
maderoi?
mapuerae = flava
oncidioides?
pachyantha
picta?
recurvata
remotiflora
selligera = granitica
spectabile = granitica
stellata
virens = diurna
wageneri = diurna

Guianas, Guyana, Suriname

affinis = guianensis
alboxanthina = pachyantha
chloroleuca
conchaechila
cordigera
dichroma?
diurna?
ensiformis? = oncidioides
granitica
guianensis
halata? = recurvata
hermentiana
ionosma
latipetala
oncidioides
pachyantha
patens?
picta
recurvata?

Brazil Including the Amazonas Area

acuta
advena
affinis = guianensis
albopurpurea
alboxanthina = pachyantha
amabile = quesneliana
amazonica
amicta
amicta subsp. *conchaechila = conchaechila*
andrichii
argentinensis
atropurpurea = randiana
bahiensis = fowliei
bicornuta = conchaechila
biflora = dichroma
bracteata
bragancae
braziliense = dichroma
buchtienii = steinbachii?
burle-marxii
cardimii
capartianum = osmantha
caximboensis
chapadensis
chloroleuca?
conchaechila
confusa
conspicua = dichroma
cordigera
cyperifolia
dichroma
doeringii
dutrai
duveenii
edithiana
ensiformis = cyperifolia
ensiformis = oncidioides
euosma = ionosma
expansa?
flava
flabellifera = ionosma
fowliei

TABLE I *(Continued)*

Brazil Including the Amazonas Area (*Continued*)

gallopavina
ghillanyi = *quesneliana*
glandulosum = *diurna?*
glutinosa = *patens*
godseffianum = *osmantha*
gonzalezii
goyazensis
granitica
guesneliana = *quesneliana*
guianensis?
hollandiae = *megalantha*
hollandiae var. *espiritusanctensis* = *megalantha*
hollandiae var. *majus?*
huebneri
hunteriana = *stellata*
icthyophylla?
insidiosa = *diota* (see Vol. V)
inversa?
ionosma
ivonae
jenischiana = *quesneliana*
latipetala
laxa = *candollei* (see Vol. V)
leucantha = *amicta?*
linearfolioides = *amicta*
longifolia = *oncidioides*
lunantha = *granitica*
lutzenbergeri
lutzenbergeri var. majus
mapiriensis = *angustiloba?*
mapuerae = *flava*
megalantha
megalantha var. *espiritusanctensis* = *spiritusanctensis*
meneziana
microtos?
microxanthina
odoratissima = *patens*
oncidioides
oncidioides var. *itabirae* = *patens*
ortgesii = *quesneliana*
osmantha
oxyphylla
pachyantha
patens
pauciflora
pedra-azulensis
pflanzii = *thrombodes*
picta?
piracanjubensis
purpurachyla = *gallopavina*
pusilla = *bracteata*
quesneliana
randiana
randii = *randiana*
recurvata?
roseum = *quesneliana*
saltensis = *argentinensis*
santos-dumontii
schmidtii
sclerocladia?
seidelii
serroniana = *patens*
spectabile = *granitica*
spiritusanctensis
steinbachii?
tarumana
thrombodes?
tripartita = *flava*
tripunctata?
unaensis
vellozoana = *oncidioides*
viridiflava
viridiflora
widgrenii?
xerophytica
xipheroides
xuxiana = *bragancae*
xuxaensis = *bragancae*
yauaperyensis

Figures 1-1 to 1-7. Diagrams of dorsal sepal, petal and lips for the South American *Encyclia* species described in this book. All sizes among the outlines are relative; actual sizes may be found in the text, with the understanding that certain variations of size between individual flowers or clones of a given species do usually occur. Drawings by Michael Ashe from type records, actual specimens of the species, or, in a few cases where these were not available, from photographs.

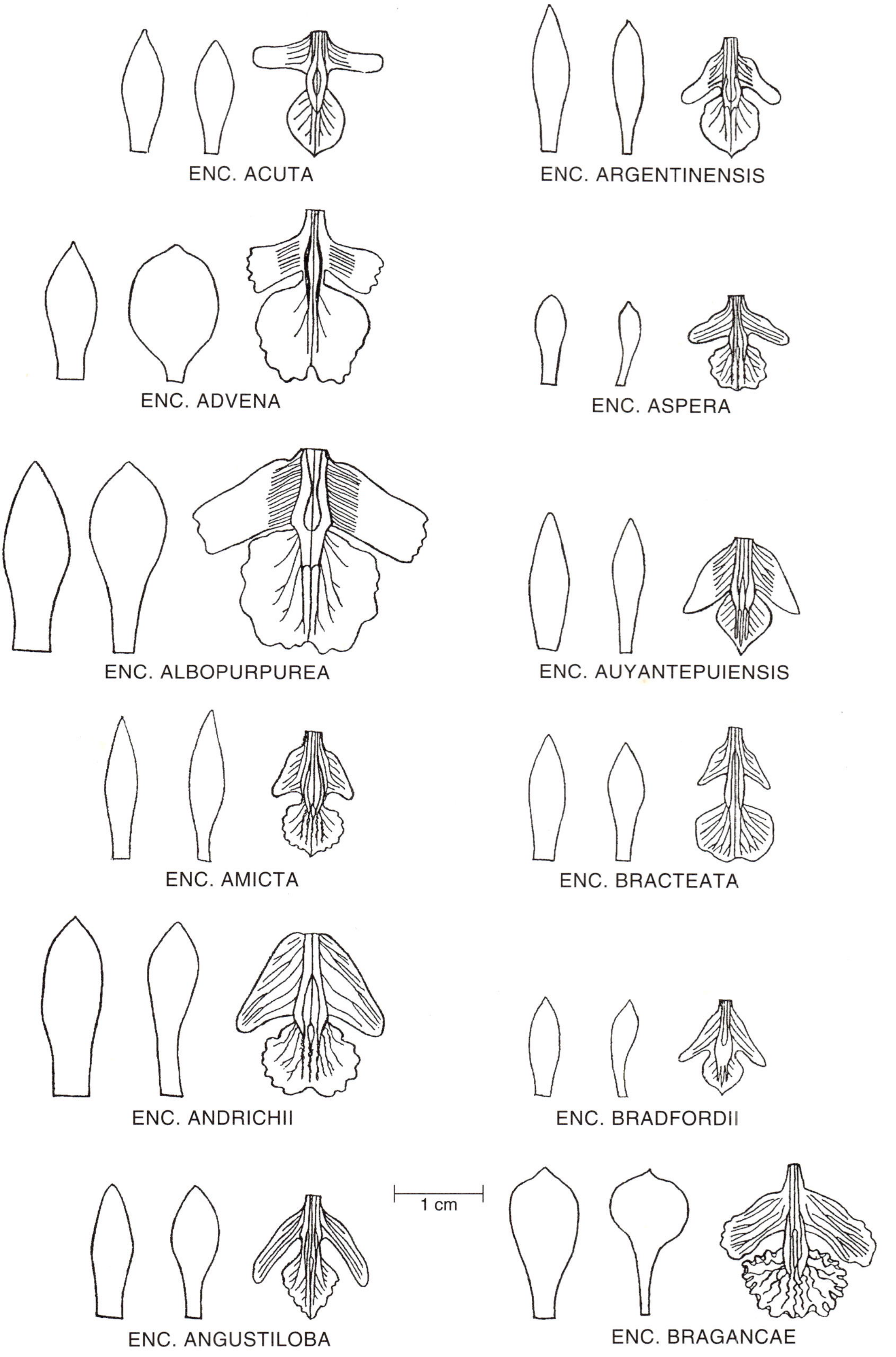

Figure 1-1. *Encyclia acuta* to *Enc. bragancae.*

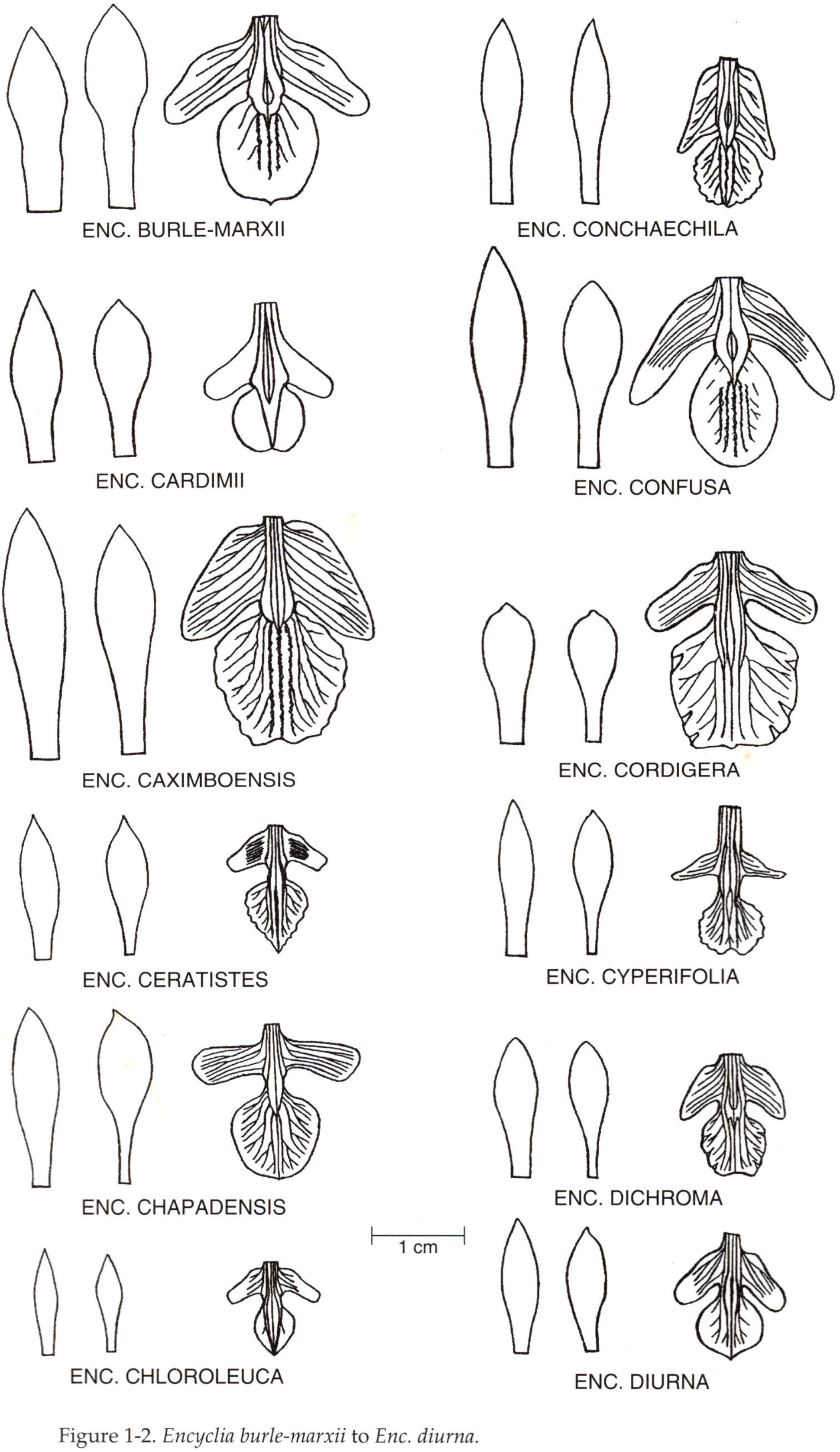

Figure 1-2. *Encyclia burle-marxii* to *Enc. diurna.*

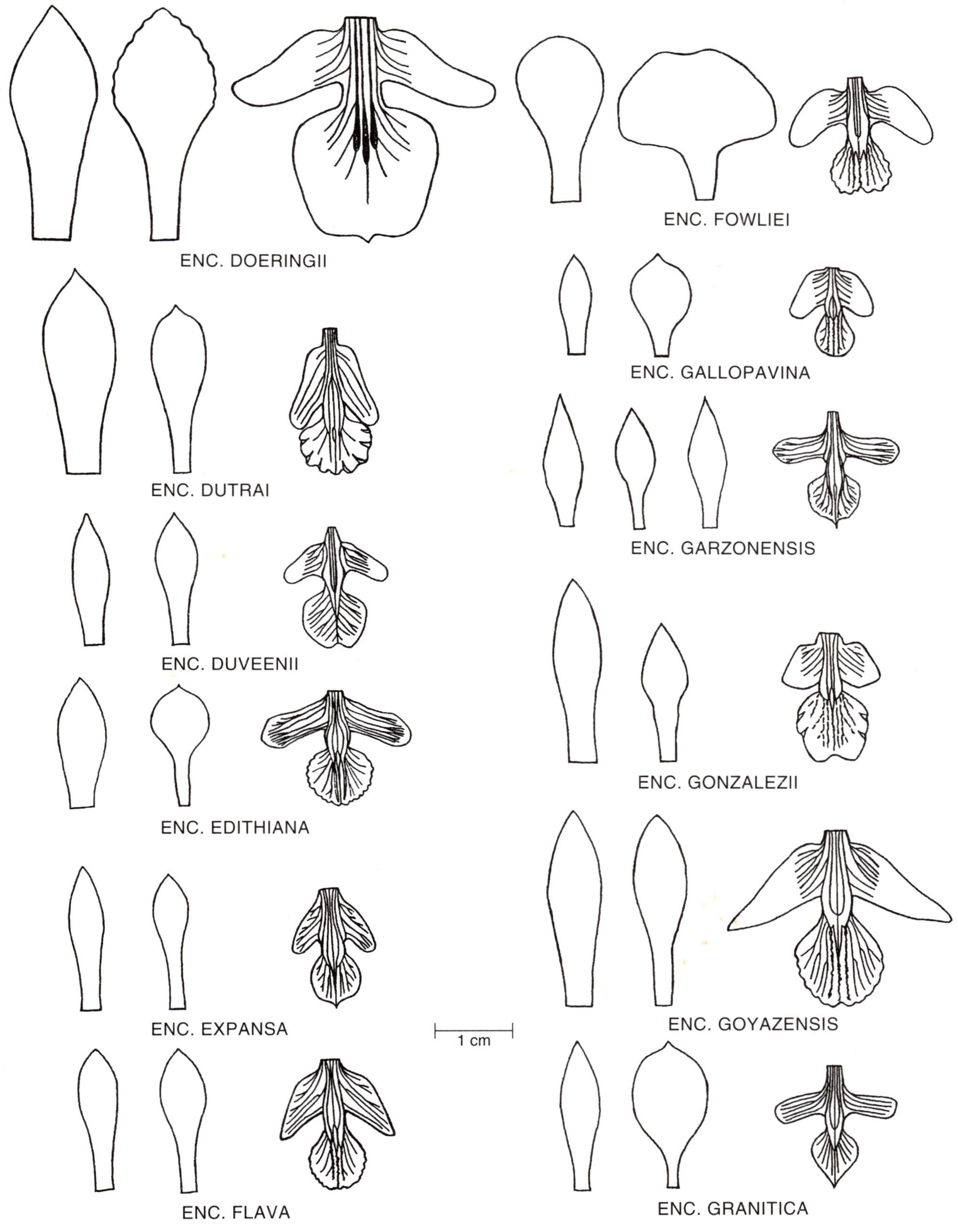

Figure 1-3. *Encyclia doeringii* to *Enc. granitica.*

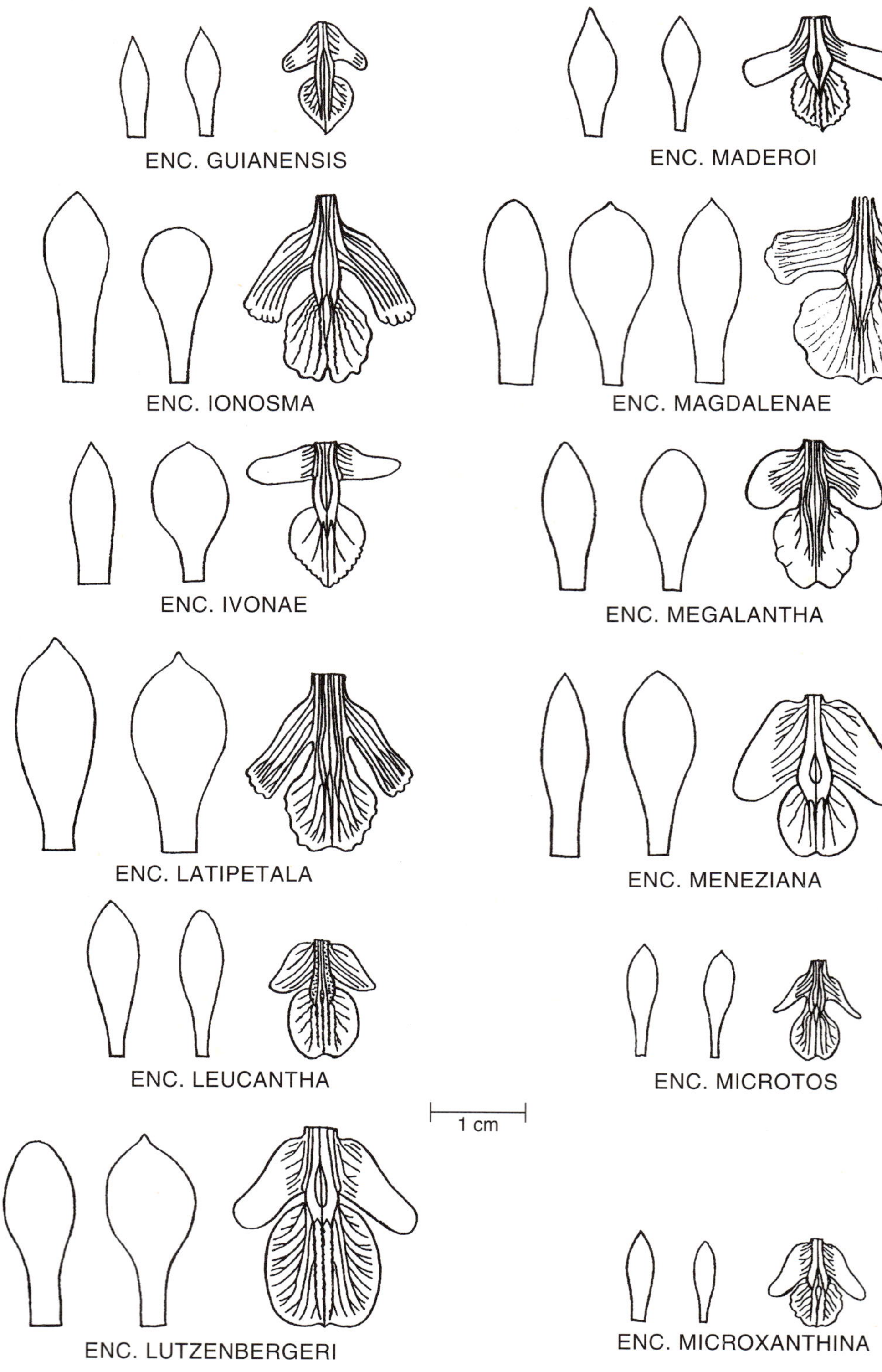

Figure 1-4. *Encyclia guianensis* to *Enc. microxanthina.*

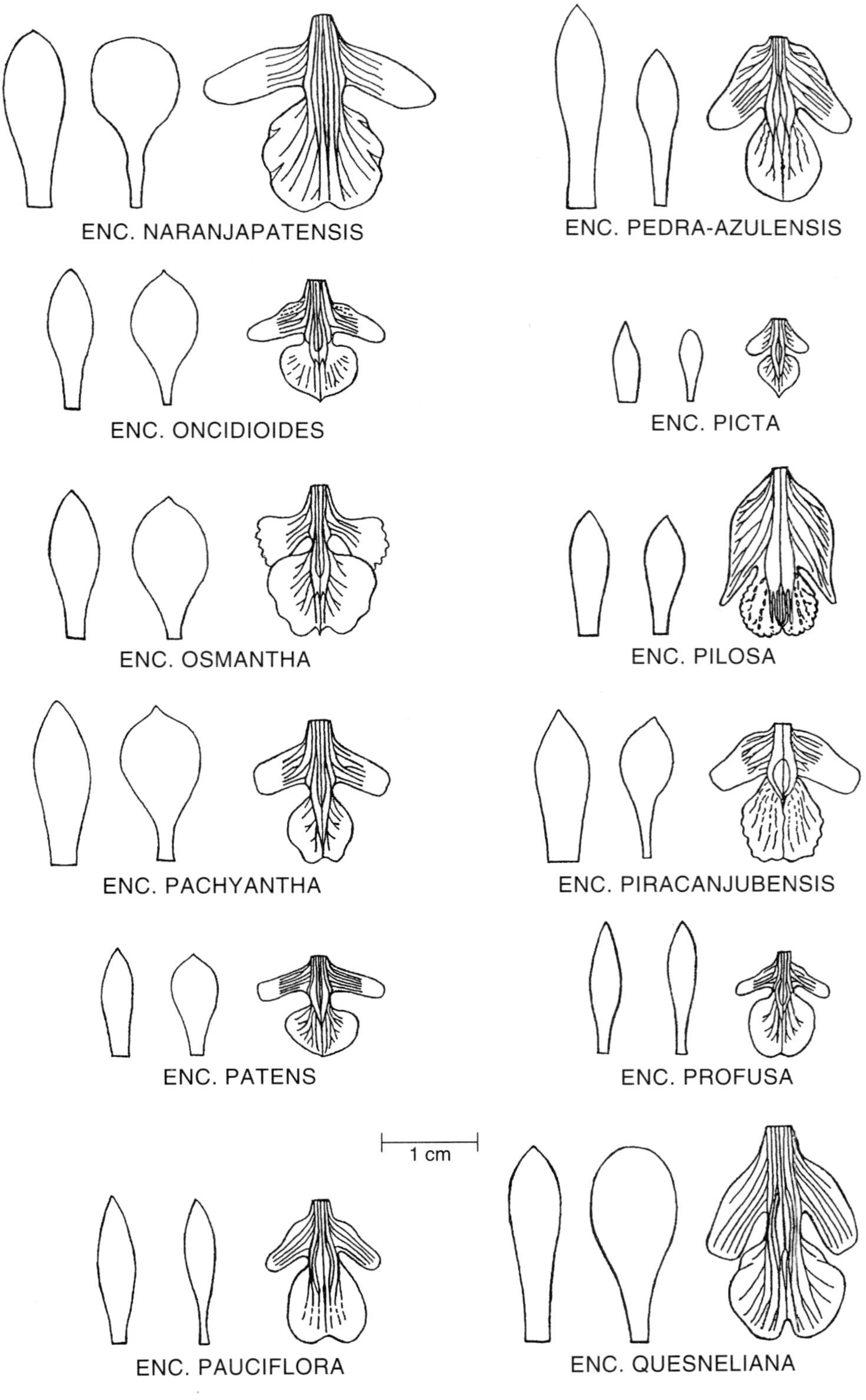

Figure 1-5. *Encyclia naranjapatensis* to *Enc. quesneliana.*

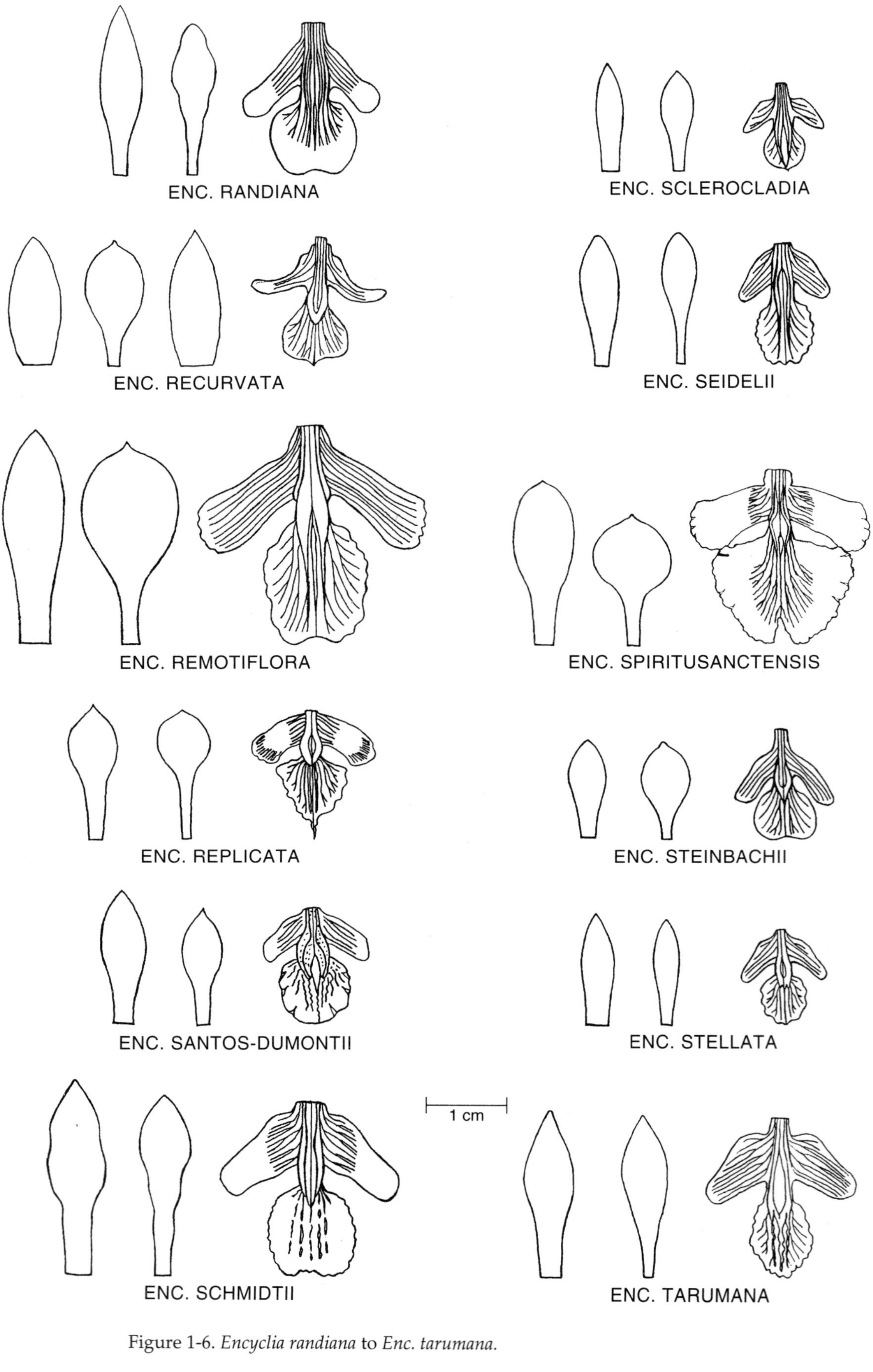

Figure 1-6. *Encyclia randiana* to *Enc. tarumana.*

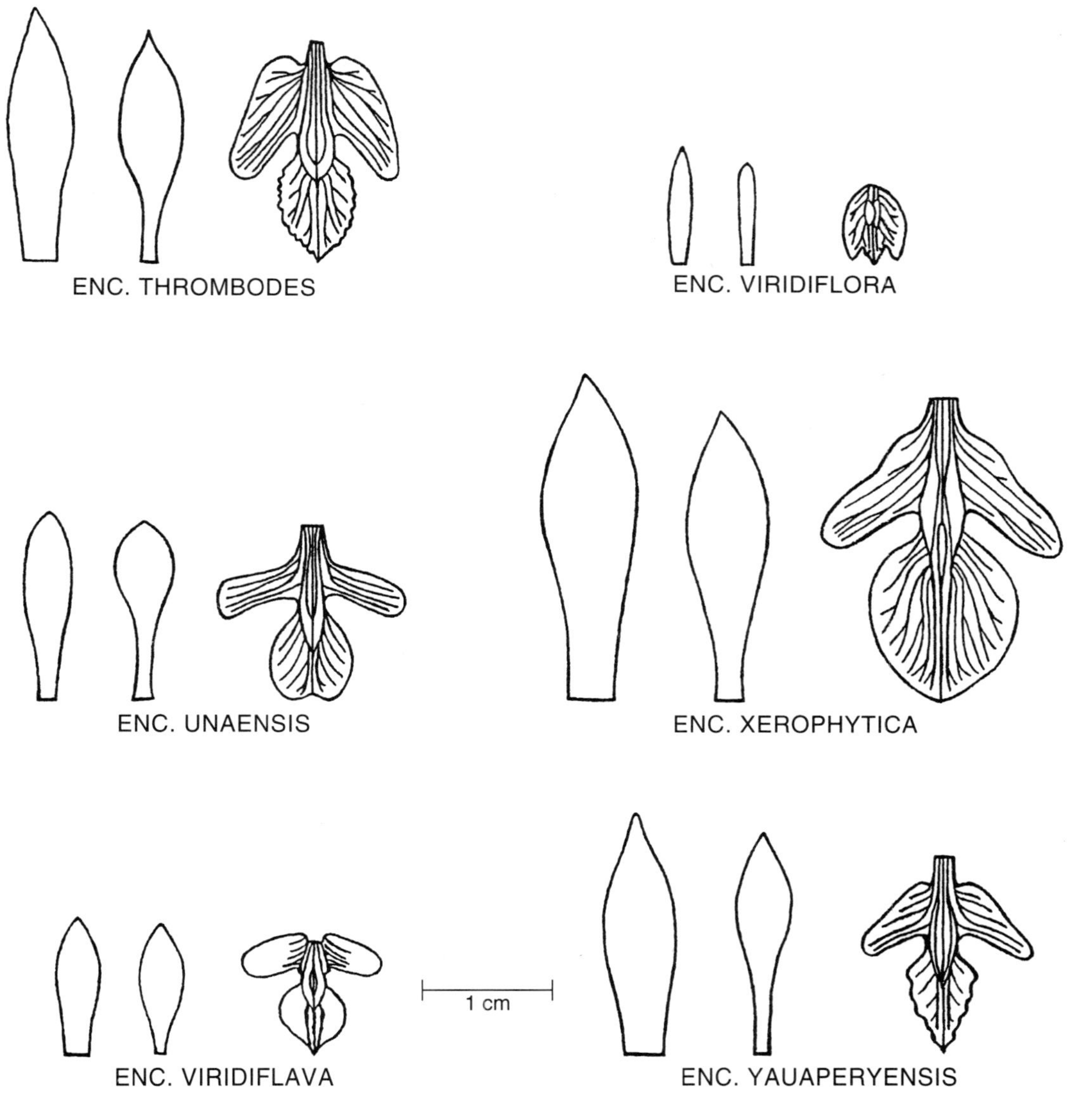

Figure 1-7. *Encyclia thrombodes* to *Enc. yauaperyensis.*

Table II

Key to Bahamian and Caribbean Species of *Encyclia*

These are dichotomous keys. Start at the beginning with a choice between the two lines marked "1." The choice of the first line marked "1" will send you to the two choices at the lines marked "2," while the choice of the second line marked "1" will send you to the pair of choices at "6." Continue making choices until your final step will, I hope, result in the name you are seeking. Comparison of the flowers with the diagrams will either confirm or negate your choice. Correct selection of choices should provide you with a proper name.

1. Flowers on an inflorescence partly or wholly cleistogamous .... go to 2
1. Flowers opening as usual, not cleistogamous .... go to 6
2. Lips without lateral lobes, or sometimes with 1 or 2 short and pointy protrusions .... *Enc. isochila*
2. Lips showing well-developed lateral lobes .... go to 3
3. Flowers in panicles with plantlets sometimes developing at the nodes .... *Enc. sintenisii*
3. Flowers usually racemose or nearly so, keikis not forming ... go to 4
4. Ovary surface rough, papillate .... *Enc. montana*
4. Ovary surface smooth, glabrous, often shiny .... go to 5
5. Midlobe of lip and lateral lobes acutely pointed ...... *Enc. bradfordii*
5. Midlobe of lip and lateral lobes rounded .... *Enc. maravalensis*
6. Plants miniature, leaves grasslike and narrow, lip without lateral lobes .... *Enc. angustifolia*
6. Plants and flowers not miniature .... go to 7
7. Rhizome twisting and turning, plant growing on palm leaf bases, in Cuba .... *Enc. ochrantha*
7. Plants more caespitose, growths in a cluster not widely separated by contorted rhizomes .... go to 8
8. Large, often reddish plants with tall pseudobulbs ruderal on coral limestone; flower stalk 3 m or more with as many as 300 flowers .... *Enc. altissima*
8. Plants and flowers otherwise .... go to 9
9. Lips of flowers boldly striped with red-purple veins; flower stalk upright; old leaves remaining on plant from non-functional abscission zone at their bases .... *Enc. gracilis*
9. Flowers and plants otherwise .... go to 10
10. Flower stalks rigidly erect with distinctly short ascending lateral branches and non-resupinate flowers .... *Enc. caicensis*
10. Flowers and plants otherwise .... go to 11
11. Plants with elongated leaves, thin pseudobulbs with bracts, and flower stalks sparsely branched; flowers with distinctly white auricles on the column and one leaf shorter than the other .... *Enc. inaguensis*
11. Flowers and plants otherwise .... go to 12
12. Flattened lip with lateral lobes as large or larger than midlobe .... go to 13
12. Lateral lobes of lip smaller than the midlobe .... go to 14
13. Flowers on widely spaced panicle, lips about 18 mm long .... *Enc. howardii*
13. Flowers with lips 8 mm long and columns short, about 3 mm .... *Enc. parviloba*

Table II *(continued)*

14. Midlobe of lip pointed at apex .... go to 15
14. Midlobe of lip rounded and sometimes notched at apex .... go to 19
15. Ovary surface distinctly papillate-echinate .... *Enc. nematocaulon*
15. Ovary surface smooth or merely asperate .... go to 16
16. Outside edges of lateral lobes of a flattened lip parallel to the axis of the lip, isthmus verrucose to pilose .... *Enc. oxypetala*
16. Lateral lobes projecting from the axis .... go to 17
17. Starry flowers yellow-brown, columns with obtuse, triangular auricles .... *Enc. acutifolia*
17. Flowers otherwise, columns without auricles .... go to 18
18. Flowers marked with a red central veining on the midlobe, lips 9–11 mm long, petals lanceolate .... *Enc. fucata*
18. Flowers with no central red veining to make a spot, lips 18–20 mm long, petals spatulate .... *Enc. selligera*
19. Flowers small to medium-sized with lip not more than 12–14 mm long .... go to 20
19. Flowers larger with lip 16–18 mm long or longer .... go to 23
20. Midlobe with large central red spot of veining surrounded by a white margin .... *Enc. tampensis*
20. Flowers otherwise .... go to 21
21. Column with wings but no auricles, lip only 9 mm long .... *Enc. kingsii*
21. Column with small auricles, lip 14–16 mm long .... go to 22
22. Sepals and petals green, lips white with 3 purple veins, and callus with 3 points onto the disc, apex rounded .... *Enc. fehlingii*
22. Sepals and petals yellow-brown, lip white with 5 short purple veins near the lip margin, apex bilobed .... *Enc. withneri*
23. Flowers nearly a concolor dull green-yellow-brown, lips reflexed into a roll and strongly arched, strongly fragrant of primroses .... *Enc. rufa*
23. Flowers not concolor with rolled arched lips .... go to 24
24. Midlobe of lip somewhat heart-shaped and cordate at base where attached to isthmus, midlobe pleated and distinctly reflexed on edges .... *Enc. plicata*
24. Midlobe rounded at isthmus, and flatter .... go to 25
25. Leaves short, stiff, pointed, usually single .... *Enc. triangulifera*
25. Leaves longer, leathery, usually diphyllous .... go to 26
26. Flowers several, pedicels short, lip violet or rose-lavender, darker veined, lateral lobes of flattened lip at right angles to the lip axis, lip to 30 mm long, strong fragrance of vanilla .... *Enc. phoenicea*
26. Flowers only a few, lip smaller, lateral lobes projecting downwards .... go to 27
27. Flowers red-brown with pale margins around sepals and petals, lip basically white but nearly covered with red-purple veining, plants particularly from the Isle of Pines, Cuba .... *Enc. pyriformis*
27. Flowers more green with red-purple veining, especially at the petal tips, plants from the main island of Cuba .... *Enc. brevifolia*

TABLE III

Key to Mesoamerican Species of *Encyclia*

1. Some or all flowers cleistogamous; ovaries somewhat swollen even at anthesis on flowers that open normally, ovaries distinctly granulate when mature, column with small auricles .... *Enc. gravida*
1. Flowers never cleistogamous, opening normally .... go to 2
2. Ovaries echinate-papillate, appearing as small gherkins when ripe .... go to 3
2. Ovaries smooth or sometimes slightly roughened, but not spiny .... go to 4
3. Lateral lobes thin and narrow, no callus on the isthmus .... *Enc. nematocaulon*
3. Lip with callus ridges, and long midlobe tapering to a narrow point .... *Enc. tuerckheimii*
4. Isthmus broad with lip tissue present on either side of the callus; sinus of lateral lobes not completely indented to isthmus .... go to 5
4. Isthmus narrow with only the callus keels making up the isthmus structure .... go to 7
5. Callus intersected by a line extended between the bases of the sinuses that separate the lateral lobes from the midlobe .... *Enc. meliosma*
5. Callus positioned above a line extended between the bases of the sinuses that separate the lateral lobes from the midlobe ...... go to 6
6. Midlobe of lip acutely pointed .... *Enc. candollei*
6. Midlobe of lip broad and fan-shaped, scarcely pointed if at all .... *Enc. flabellata*
7. Lateral lobe apices enlarged and well rounded, wider than mid-portions, no column projections .... go to 8
7. Lateral lobe apices tapered, pointed, rounded, but not enlarged, column various .... go to 14
8. White lip with band of violet-red at the base of each lateral lobe and around the disc below the callus .... *Enc. mooreana*
8. Coloration otherwise .... go to 9
9. Flowers large, lips 30–45 mm long, lip colors white to pink or rose .. .. *Enc. cordigera*
9. Flowers mid-sized or smaller, lip 12–25 mm long .... go to 10
10. Lip 20–25 mm long, sepals and petals dull yellow-brown with wavy margins, veined splotchily with red-brown, lip adnate to column 1–2 mm, column bowed, lip white with rosy flush and completely veined in red-purple .... *Enc. hanburyi*
10. Flowers otherwise .... go to 11
11. Flowers with rich red-brown shiny petals and sepals with maroon-red lip about 13 mm long, 17 mm in spread across flattened lateral lobes .... *Enc. atrorubens*
11. Flowers otherwise .... go to 12
12. Sepals and petals tan-brown with reddish veins and thin yellow margins; apices of lateral lobes turned back like ears with red-purple veining and a yellow margin .... *Enc. diota*
12. Flowers otherwise .... go to 13
13. Disc of lip without colored veining, prominent green sepaline nectaries, flower stalks scarcely, if at all, above leaves ... *Enc. suaveolens*
13. Flower stalks wide-spread panicle or long raceme ..... *Enc. selligera*
14. Lateral lobes of lips small and triangular in shape, spread the same or less than the width of the midlobe on a flattened lip .... go to 15
14. Lateral lobes of lip larger, not so triangular, and the spread of the tips on a flattened lip is wider than the midlobe .... go to 15

TABLE III *(continued)*

15. Lip white and wine-purple, midlobe curled down on the midline, the columns with auricles, flowers in tall panicles .... *Enc. pollardiana*
15. Lip white and yellow attached 1–2 mm to the base of the column, veined on the midlobe with red-purple, callus puberulent, flowers mostly in racemes, in warm, dry lowland areas ..... *Enc. trachycarpa*
16. Lateral lobes with upper edges parallel to lip axis with only the tips flared .... *Enc. microbulbon*
16. Lateral lobes otherwise .... go to 17
17. Lower edge of lateral lobes projecting at right angles to lip axis .... go to 18
17. Lower edge of lateral lobes at an angle of 40–50 degrees from the lip axis .... go to 19
18. Flowers with brown sepals and petals, lips 11–12 mm long, puberulent at base, pronounced brown area at base of lateral lobes .... *Enc. chiapasensis*
18. Flowers with deep red-brown sepals and petals with glistening surface; column with auricles; lip white with a green flush at the base .. .. *Enc. dickinsoniana*
19. Isthmus area 2–3 times longer than wide .... *Enc. bractescens*
19. Isthmus not much longer than wide .... go to 20
20. Columns without auricles, wings, or appendages .... go to 21
20. Columns with auricles, wings, or teeth .... go to 25
21 Lip pointed, acutely or obtusely .... go to 22
21. Lip rounded or notched at apex, sometimes with apicule ... go to 24
22. Flowers yellow-brown in 1 m spreading panicles, lip creamy with purple veining, sepals and petals oblanceolate with acute tips and paler yellow edges .... *Enc. cepiformis*
22. Different from above .... go to 23
23. Flowers green-brown with purple veining on white lip midlobes, ovary surface smooth .... *Enc. peraltensis*
23. Flowers tawny yellow with fine purple veining on white lip midlobes, ovary surface smooth .... *Enc. oestlundii*
24. Flowers thin textured, pale green to light tan, with verrucose veining on the lip and occurring in a densely flowered, nodding panicle; lip apex varying from somewhat rounded to almost pointed .... *Enc. incumbens*
24. Flowers with broad, exposed pale yellow to white column with small lateral lobes projecting on either side; base of lip attached to column for 2–3 mm, with strong central rib, sepals and petals brownish yellow .... *Enc. adenocarpon*
25. Flowers with pointed lips .... go to 26
25. Flowers with rounded lips, sometimes notched at the apex with an apicule .... go to 33
26. Column with thin, hornlike appendages at the tip; oval brown patches at the base of the lateral lobes, green flowers with white lips marked with veins .... *Enc. ceratistes*
26. Columns with wings, toothlike protrusions, or auricles ..... go to 27
27. Column with distinct auricles .... go to 28
27. Column with only winged edges, sometimes forming a tooth near the tip .... go to 29
28. Column about 7 mm long, 3 central main veins extending from callus onto disc of midlobe, ovary surface smooth .... *Enc. amanda*
28. Column about 10 mm long, central vein forming a rib to the apex of the midlobe, ovary surface rough, flower stalk only to leaf height ... . *Enc. asperula*

TABLE III *(continued)*

29. Raised central vein extended to a mucronate tip on the midlobe, white column, anther cap, and midlobe, band of purple veins on the midlobe, and lateral lobes green-yellow with brown veining on the reflexed tips . . . . *Enc. aenicta*
29. Central vein not so distinct, lip color otherwise . . . . go to 30
30. Lateral lobes of lip 7–8 mm long, narrow, with parallel sides and with tip crenate or scalloped . . . . *Enc. perplexa*
30. Lateral lobes shorter, broader, with rounded apices not scalloped . . . . go to 31
31. Anther cap and tip of column bright red, sepals and petals bright green to dull green-brown, lateral lobes equally colored, midlobe white without markings . . . . *Enc. lorata*
31. Anther cap and tip of column not red . . . . go to 32
32. Lip adnate to column for 2–3 mm, peduncles of flowers papillose, midlobe scarcely veined with purple and ridged, creamy yellow . . . . *Enc. papillosa*
32. Lip attached to column with unusual adnation, 3–5 red-purple verrucose veins on white midlobe, ovary surface rough, warty . . . . *Enc. spatella*
33. Lip more or less white, about twice the length of the column with 5 major veins faintly marked with purple, sepals and petals dull olive green . . . . *Enc. stellata*
33. Lips otherwise . . . . go to 34
34. Sepals and petals green at base, usually shading toward brown tips with maroon veining, column and lip pale yellow, richly veined on the midlobe in dark red, front edge of lip with clear deep yellow to orange ruffled border, lips about 20 mm long . . . . *Enc. alata*
34. Lips without a clear ruffled gold-yellow front edge . . . . go to 35
35. Flowers concolor rose-magenta; ovary surface slightly rough; column bowed upwardly; midlobe broadly shovel-shaped; lateral lobes tapered to a dull point . . . . *Enc. kennedyi*
35. Flowers otherwise . . . . go to 36
36. Flowers pale pink-lavender with darker veining on lip midlobe, midlobe about twice the width in length with the callus reduced to parallel keels . . . . *Enc. adenocaula*
36. Flowers otherwise, not rosy or pink-lavender . . . . go to 37
37. Medium-sized flowers with sepals and petals to 21–23 mm long, midlobes of lips wider than long . . . . go to 38
37. Flowers larger with midlobes longer than wide . . . . go to 39
38. Lips white with a few purple veins on disc, columns white, sepals and petals dull olive with brownish veining, petals usually more erect and closer to the dorsal sepal . . . . *Enc. guatemalensis*
38. Lips yellow covered with red-brown verrucose veining, including the lateral lobes, sepals and petals green to deep coffee brown, petals in usual position . . . . *Enc. parviflora*
39. Petals obovate-spatulate, tan-brown with green bases, lip creamy and much covered with verrucose maroon veins and spots . . . . *Enc. ambigua*
39. Sepals and petals dull yellow green, lip white with verrucose red markings . . . . *Enc. virens*

TABLE IV

Partial Key to South American Species of *Encyclia*

1. Flowers with "black" anther caps that are bilobed; also callus area at base of lip hairy or pubescent (hand lens), sometimes finely so, and with ovary surface and flower stalk usually rough or asperous ...... go to 2
1. Flowers otherwise ..... go to 6
2. Lip 12 mm long, midlobe longer than wide, pointed, white turning yellow, lateral lobes with maroon veining over green and with apices white; sepals and petals green; ovary surface smooth ..... *Enc. thrombodes*
2. Flowers otherwise ..... go to 3
3. Flower stalks shorter than the narrow, pointed leaves; lip 15–17 mm long with ridges like ribs on a scallop shell, lateral lobes pointed; column straight with winged edges and no auricles; ovary surface rough ..... *Enc. conchaechila*
3. Flowers otherwise ..... go to 4
4. Lip with unique hairy callus pad, midlobe rounded with notched apex and 3–5 main verrucose veins; ovary surface rough; column wings folding in at tips ..... *Enc. leucantha*
4. Flowers different ..... go to 5
5. Lateral lobes of lip triangular, stubby, with rounded tips, midlobe rounded with pointed apex and 5 main veins, the central vein keel-like ..... *Enc. amicta*
5. Smallish plants 23 mm high with stout roots and single, narrow leaves; flowers small, mostly racemose, with sepals and petals yellow-green with brown overtones; lateral lobes of lip long and droopy from a wide triangular base to extend to a level halfway down the midlobe; unique hairy mat of hairs (pilose) with 5 verrucose veins on the disc with the central vein ending in an apiculus; the arcuate column is slightly winged ..... *Enc. pilosa*
6. Epiphytes with single, pendent, thin, narrow leaves from marble-sized pseudobulbs; flower stalks with only 1–2 flowers each; sepals and petals green mottled with brown; lip a bright magenta ..... *Enc. bracteata*
6. Flowers and plants otherwise ..... go to 7
7. Flowers unique, fragrant, and frizzled, with frilly pink midlobes and lateral lobe tips that are crisped and even fimbriate; lateral lobes yellowish with purple veining; petals spatulate and clawed; column yellow with small auricles ..... *Enc. bragancae*
7. Flowers different from above ..... go to 8
8. Flowers ordinarily with rosy magenta sepals and petals and darker lips ..... go to 9
8. Flowers in varied combinations of brown, green, yellow, or white, any rosy color confined to the lip ..... go to 10
9. Shouldered lip with triangular lateral lobes flaring toward their obtuse apices; lip 21–24 mm long; column 10 mm long with distinct auricles ..... *Enc. dichroma*
9. Tapered, clawed lip base, lateral lobes oblong with truncate apices; lip about 25–27 mm long ..... *Enc. quesneliana*
10. Flowers yellow-brown with unique dark brown tessellations; very wide petals, rotund and shortly clawed, paddle-shaped; lip 14 mm long; column white, with auricles ..... *Enc. fowliei*
10. Flowers otherwise ..... go to 11

TABLE IV *(continued)*

11. Flowers with smallish lateral lobes of unusual rhombic or diamond shape; lip with slightly clawed base; sepals and petals oblanceolate, greenish with brownish veining; lip midlobe white to yellow with purple veining; column 7 mm and white .... *Enc. gonzalezii*
11. Not as above .... go to 12
12. Flowers green with sepals speckled in puce-maroon and with a narrow yellow edge; lip whitish to creamy yellow with a small crescent of purple veins on disc, midlobe rounded with wavy margin, a distinct point and with a central rib; column green with purple marking, with auricles; ovary surface smooth .... *Enc. granitica*
12. Different .... go to 13
13. Small, non-resupinate flowers with lateral lobes fused to midlobe so that their apices and the pointed midlobe are all 3 of equal size and in a line; petals ligulate-lanceolate; column only 5–6 mm long, somewhat winged, with rounded protrusion near the apex ...... *Enc. viridiflora*
13. Otherwise .... go to 14
14. Flattened lip with lower edge of lateral lobes at right angles to the lip axis .... go to 15
14. Flattened lip with lateral lobes not at right angles, but pointing downwardly .... go to 19
15. Lateral lobes pointed from a short triangular base attached below the lip claw of about 3 mm; sepals and petals yellow tinted with pale rose; anther cap bright yellow; column curved upwardly .... *Enc. cyperifolia*
15. Lateral lobes oblong and rounded .... go to 16
16. Lip midlobe rounded with or without a small apiculus at the tip of the midrib .... go to 17
16. Lip midlobe broadly obcuneate or shovel-shaped coming to a point .... go to 18
17. Thin-tissued flowers; sepals and petals green-tan and reflexed toward ovary; wide callus extending to white midlobe, apex with the 2 halves of the midlobe reflexing along the central line; flowers in wide panicles .... *Enc. garzonensis*
17. Flower parts thicker textured, not reflexed; sepals and petals red-brown, lip purple veined with a white margin; flowers in tall racemes to 1 m or more .... *Enc. chapadensis*
18. Sepals and petals obovate to oblanceolate, green with brown flush, lip white with 1–3 purple veins down center, about 14 mm long; column with auricles .... *Enc. acuta*
18. Dorsal sepal and petals unlike, the petals about twice the width of the sepal and obovate-spatulate with a short, broad claw .... *Enc. ivonae*
19. Fragrant; flattened lip showing lateral lobes with tips curling upward and clear of veining; sepals reflexed backward; sepals and petals green, lip creamy with darkened veins .... *Enc. recurvata*
19. Flowers otherwise .... go to 20
20. Slightly clawed lip with lateral lobes tapering to a sharp point; 6–7 major veins on midlobe verrucose; ovary surface and flower stalks well roughened, stalks reddish, anther cap may be dark maroon spotted; auricles on column pointlike extensions .... *Enc. microtos*
20. Not like above .... go to 21
21. Flowers small, not medium sized or large; lips usually not longer than 14–15 mm in length .... go to 22
21. Flowers with larger lips, medium sized to large; lips mostly measuring more than 15 mm in length to as much as 30–40 mm .... go to 39
22. Plant with some or all flowers cleistogamous, normal or dissected flowers with a wide isthmus and pointed lateral lobes and obovate pointed midlobe .... *Enc. bradfordii*
22. Flowers not cleistogamous and different from above .... go to 23

TABLE IV *(continued)*

23. Sepals and petals green, oblanceolate; in arching panicles; lip white, obcuneate and pointed with purple veining; tips of lateral lobes reflexing sharply at column end, and having a distinctive oval brown spot at the base of each lobe when pulled open; column with a fine tooth or hornlike process on each side near the apex .... *Enc. ceratistes*
23. Flowers not the same .... go to 24
24. Plants with 2–3 mm wide coarse roots and leaves with thick, stiff, shiny leaves with sharp tips and a twist; flower stalks 12–13 cm tall with only 3–5 flowers 28 mm across; lip green at base, but mostly an intense yellow with a white callus; veining verruculose and anther cap dark maroon .... *Enc. microxanthina*
24. Otherwise .... go to 25
25. Flower stalk and branches coarse and asperous; flowers small with sepals and petals 12 mm long, green with maroon veining; lip shorter with green lateral lobes and yellow midlobe; column only 6 mm long .... *Enc. sclerocladia*
25. Flowers not so .... go to 26
26. Lip with rounded midlobe and rounded apices, with or without an indentation at the apex, and with or without an apiculus ... go to 27
26. Lip with obovate-cuneate shovel-like shape tapering to a narrowed or obtuse point .... go to 36
27. Lip midlobe diamond shaped, ruffle-edged, with central vein that runs to the apex; lateral lobes unusually long and narrow; sepals and petals red-brown; lip white; asperous veining; ovary surface asperous .... *Enc. angustiloba*
27. Not anything above .... go to 28
28. Pseudobulbs top-shaped; foliage short; racemes distinctive; sepals and petals green; lip white; column green, with auricles and a yellow anther cap; lateral lobes short, oblong, and with rounded tips; midlobe shovel-shaped with the callus ridges continued nearly to its tip .... *Enc. guianensis*
28. We're not finished .... go to 29
29. Flowers on nodding panicles; sepals and petals brown bordered in yellow with a white lip veined in rose; lip halves folded down; lateral lobes of lip oblongish with rounded tips and apical veining; midlobe with a somewhat elongated acuminate tip from triangular disc; column with auricles; ovary surface smooth .... *Enc. replicata*
29. Midlobe rhombic or diamond-shaped, longer than wide, with a small rounded apex; lateral lobes triangular with rounded apices; sepals and petals olive green, midlobe white with some reddish streaking; lateral lobes and column creamy with red veins; white callus ridges with red markings .... *Enc. yauaperyensis*
30. Small plants with 15 cm leaves; lip 12–13 mm long with lateral lobes a trapezoidal shape with rounded tips and as large or larger than the midlobe; flowers small in 60 cm panicles; spatulate petals 15 mm long with broad claw; midlobe with central purple-veined area .... *Enc. gallopavina*
30. Not the above combination .... go to 31
31. Larger plants and flowers, leaves to 30 cm, long and narrow; sepals and petals oblanceolate; lip midlobe with 5 raised veins, pointed lateral lobes; column narrow, slightly winged .... *Enc. flava*
31. Not above .... go to 32
32. Sepals and petals dingy green, fleshy, in a short dense flower panicle little more than leaf height; lip 12–13 mm long, creamy with the midvein purple-red; midlobe broadly obovate-cuneate, wider than long; column white and with auricles; intensely fragrant ...... *Enc. patens*
32. Not so .... go to 33

TABLE IV *(continued)*

33. Sepals and petals dull yellow with red-brown veining and a thin texture; plants with flowers grow to 40 cm; lip 14 mm long, column 8–9 mm, slightly curved .... *Enc. picta*
33. Lips with large, fleshy callus ridges extending over the disc and forming a rib to the lip apex .... go to 34
34. Small plants with 5–7 flowers on fractiflex stalk; flowers lime-green, firm and fleshy; lip with cordate base and 10 mm long; column only 5 mm long and with auricles; ovary surface rough ..... *Enc. viridiflava*
34. Otherwise .... go to 35
35. Sepals and petals fleshy green-yellow; lip 15 mm long with white callus and a purple midvein; column dark colored with auricles and bright yellow anther cap; ovary surface smooth ..... *Enc. chlorolueca*
35. Otherwise .... go to 36
36. Plants with smallish pseudobulbs and somewhat narrowed leaves to 24 cm long; dorsal sepal oblanceolate and petals subspatulate, green-brown with a fine yellow border; lip about 11 mm long, white with bright magenta veining forming a spot on the disc; midlobe oval from a tapered base and indented at the apex .... *Enc. unaensis*
36. Otherwise .... go to 37
37. Medium-sized plants with tall panicles of 15–20 flowers and 2–3 leaves; sepals and petals greenish with brownish tips; lip white with rosy tints and purple veining; lateral lobes greenish and veined brown; narrow white margin around the lip; midlobe circular with verrucose veining and a ruffled edge with a slight notch at the tip; lateral lobes triangular from a broad base and with rounded apices .... *Enc. andrichii*
37. Otherwise .... go to 38
38. Sepals and petals oblanceolate and green-brown; lip white with red-purple veining; midlobe rounded with an apiculus; fleshy callus ridges form a wide isthmus and the central vein continues to the apex; column without auricles .... *Enc. expansa*
38. Sepals and petals green-brown, lip white; column green with wingy edges that dip down to form a blunt projection on either side of the tip; lip midlobe rounded with slightly scalloped edges and 5 raised veins; veins show a few hairs (hand lens) .... *Enc. stellata*
39. Flowers with large lips, more than 25 mm long .... go to 40
39. Flowers with medium-sized lips, less than 25 mm long but more than 15 mm .... go to 46
40. Sepals and petals rich red-brown, lip white with bold, large central red spot on the disc; lateral lobes tipped in white, in large broad panicles to 100 flowers; petals obovate with long, narrowing base and undulate edges .... *Enc. randiana*
40. Not as above .... go to 41
41. Lip heavy shouldered with oblique lateral lobes with rounded white apices; isthmus narrow but supporting a rounded white midlobe with wavy edges and notched apex; midlobe covered with fine reddish veining; column white, 13 mm long; flowers in a long, nodding panicle .... *Enc. megalantha*
41. Not a match .... go to 42
42. Lip with 2 mm claw; squarish midlobe about 18 mm across the base and 19 mm long, apiculate, truncate; isthmus 5 mm broad and 3 mm long; lateral lobes spreading, with rounded tips, 15 mm long, and shouldered at their base; 3 central veins ridged and thickened over the disc of the midlobe; petals broadly obovate tapering to a wide claw, wavy edged .... *Enc. doeringii*
42. Otherwise .... go to 43

TABLE IV *(continued)*

43. Sepals curled forward at tips with prominent green nectaries; sepals and petals varying in color from green to brown; lip also variable from white with some red veining to concolor pink or rosy magenta; no callus enlargements, but 5 central veins toward the broad truncate apex of a large midlobe; lips to 4 cm wide and 5 cm long .... *Enc. cordigera*
43. It is ahead .... go to 44
44. Plants with 2–3 lineal acute leaves and panicles 3 times the lip length with flowers smelling of vanilla; sepals and petals yellow-brown, lip white, large, with carmine veining; column winged toward apex but auricles lacking; sepals and petals a rich brown, obovate-oblanceolate; lip midlobe reflexed on midline with 3–5 main veins; lateral lobes oblong with square scalloped tips and veined only at the bases .... *Enc. albopurpurea*
44. Otherwise .... go to 45
45. Flowers large with oblanceolate-obovate dorsal sepal and nearly orbicular petals with short claws; sepals and petals olive green with brownish veining and flushes; lateral lobes oblong with expanded apices without darker veining and greenish bases; midlobe creamy white with purple veining and retuse; lip has a 3 mm claw and a thin narrow isthmus with callus ridges ending on upper portion of disc .... *Enc. advena*
45. Little different from above, but flowers more fleshy and cupped in form; sepals and petals brown and yellow-green; petals obovate; lips white with carmine veining; lateral lobes expanded and crenate; midlobe broad, retuse and with an apiculus; callus ridges extending halfway down disc of lip .... *Enc. osmantha*
46. Growths separated by distinct, easily visible rhizomes at least 2–3 cm long between growths; plants sometimes growing on rocks or terrestrially in white sand areas .... go to 47
46. Growths not separated by easily seen rhizomes; plants usually epiphytic; pseudobulbs in clumps or clusters .... go to 50
47. Large terrestrial or ruderal plants with pseudobulbs to 9 cm tall with 3 to 4 leaves 37 cm long; growths well separated by several centimeters; panicle stoutly upright to 90 cm tall, long stalked with branches densely arranged and flowered; flowers blotchy tawny orange-brown but not spotted as such; petals spatulate, yellow edged, clawed; lip white to creamy with crescent of red-purple veins, broadly shovel-shaped and pointed; sweetly fragrant like roses or violets .... *Enc. oncidioides*
47. Proceed .... go to 48
48. Plants smaller than in 37, mostly on rocks with short rhizomes between growths with 2 leaves; thin erect stalks and well-spaced starry flowers; "black horned" anther cap, but according to the type description, a smooth callus and ovary surface; lip lateral lobes triangular and pointed; 3 keeled veins as main callus ridges; sepals and petals purple-brown with yellow borders, lip midlobe white with purple veining .... *Enc. auyantepuiensis*
48. Otherwise .... go to 49

TABLE IV *(continued)*

49. Bifoliate growths mostly with 3 cm rhizomes; flower stalk smooth with 10–17 flowers on short branches; lip 18–19 mm; lateral lobes obliquely triangular, blunt; central vein of midlobe thickened at the base between the 2 main callus ridges and running to the apex .... *Enc. pedra-azulensis*
49. Plants epiphytic producing large clumps with growths about 3 cm apart, leaves erect, linear; smooth flower stalk with a few branches may reach 60 cm; lateral lobes of lip shouldered, bluntly triangular, oblique; midlobe rounded with wavy margin and 3 major veins on the disc; sepals and petals bright green at the base, brown toward the tips; lip white with lilac veinings at the base .... *Enc. tarumana*
50. Verrucose flower stalk with many reddish papillae and verrucose ovary surface; flowers fragrant of raspberries or cherries, 5–7 on each fractiflex branch of panicle; sepals and petals a dull ochre-yellow, petals spatulate-clawed, lip 16 mm long; column 5–6 mm, winged, with yellow auricles; anther cap spotted with red speckles .... *Enc. aspera*
50. Not as above .... go to 51
51. Sepals and petals a glaucous red-brown with thin green edge; lip with narrow whitish margin, 19 mm long, with 3 mm claw and 2 mm isthmus; lateral lobes clear of veining; column with auricles; ovary surface rough .... *Enc. magdalenae*
51. Otherwise .... go to 52
52. Plants with 3 linear, oblong, rigid leaves and erect racemes; flowers red-brown with white lip 13 mm long with lateral lobes yellow, veining purple; ovary surface smooth .... *Enc. burle-marxii*
52. Continue below .... go to 53
53. Plants with 2 semi-upright, stiff leaves; flower stalks shorter than leaves and with 2–3 flowers; sepals and petals green suffused purple; lip white with red-brown veins; lip lateral lobes with long wide base, shouldered, downswept, and rounded; lip midlobe somewhat pleated with scalloped edge, transversely oval and reflexed on midline .... *Enc. dutrai*
53. Not as above .... go to 54
54. Plants with small, fusiform pseudobulbs, 3 narrow leaves to 34 cm long, and few flowers on short stalks at leaf height; sepals and petals green-brown; lip 16 mm; column 10 mm; lip midlobe white with purple veining, wavy edged .... *Enc. santos-dumontii*
54. Not yet described above; plants with green-brown-yellowish sepals and petals, white lips; species difficult to distinguish; please use the flower outline drawings as you go as an aid in identification .... go to 55
55. Leaves 1–2; flowers in long 50 cm racemes with only a few flowers; at least 5 purple verrucose veins on the white lip that is longer than wide; lateral lobes long and winglike, pointed; column yellow-green, curved .... *Enc. confusa*
55. Otherwise .... go to 56
56. Sepals and petals yellow-green with brown veining and spots toward apices; flowers fragrant; lip white with rose-pink veining and white edge; column 7 mm long, with auricles .... *Enc. edithiana*
56. Keep going .... go to 57
57. Large plants with panicles of 20–25 flowers; sepals and petals green-brown; long winglike pointed lateral lobes with unveined tips; white callus with rosy veins; wing points on column .... *Enc. goyazensis*
57. Otherwise .... go to 58

TABLE IV *(continued)*

58. Bifoliate plants with erect panicle to 60 cm; sepals and petals green with darker veins and yellow edges; petals spatulate; midlobe rounded with 3 purple veins; lip lateral lobes truncate with rosy tips; midlobe pinkish with white edges .... *Enc. lutzenbergeri*
58. Otherwise .... go to 59
59. Epiphytic plants with green-brown pseudobulbs to 5 cm high and 1–2 leaves to 40 cm long; erect raceme to 60 cm with 6–10 flowers; sepals and petals green-brown; lip white, sometimes flushed purple and with purple veining; callus ridges with some veining and spotting; lateral lobes shouldered, wide, and almost square at the tips; midlobe rounded from a broad base with slightly scalloped down edges; column green with white apex 12 mm long and slightly curved; anther cap yellow .... *Enc. piracanjubensis*
59. Not any of the above .... go to 60
60. Mostly racemose flowers with oval lip midlobes that are longer than broad and with ruffled edges; basinlike callus on isthmus with 3 teeth extending into veins on the disc; lateral lobes ligulate with rounded apices; sepals and petals a dull greenish yellow with maroon markings and midlobe white with purple veining; column with auricles; ovary surface rough .... *Enc. argentinensis*
60. Otherwise .... go to 61
61. Plants with conical pseudobulbs and 3 straight, fleshy leaves; racemes with 10 flowers; sepals and petals olive green; lip white with purple veining; lip lateral lobes narrowly ovate, obliquely arranged; midlobe rounded to a blunt point; column 12 mm long, with auricles; ovary surface rough .... *Enc. xerophytica*
61. More to come .... go to 62
62. Plants with brown pseudobulbs, 2–3 leaves, and a flower stalk that may rise to 50 cm with only a few fragrant flowers; petals distinctly shorter than the dorsal sepal; sepals greenish with brown veining; lip white from a green base and covered with purple veining even on the callus; column white suffused with reddish stains .... *Enc. caximboensis*
62. Otherwise .... go to 63
63. Sepals and petals tawny brown; callus wide with a rib extending to the lip apex; plants small with racemes 15 cm tall and having 4 flowers; base of the lip a tapering claw giving rise to ligulate, rounded lateral lobes; midlobe circular, slightly indented, with an apiculus; ovary surface slightly rough .... *Enc. cardimii*
63. Otherwise .... go to 64
64. Wide, long panicles of starry bright green flowers with white lips and a few short slightly raised purple lines on the midlobe; midlobe circular with an apiculus; lateral lobes oblong-ligulate and rounded; column 10 mm, with small auricles, green, with yellow anther cap .... *Enc. diurna*
64. Try another species .... go to 65
65. Sepals and petals dark brown, reflexing back alongside the ovary; pseudobulbs with 2 leaves; lip an intense rosy magenta; panicle sparsely branched, 50 cm long, with 20 flowers; lip with long isthmus, a circular midlobe that is notched at the end, and slightly raised veins; column 10 mm long, with auricles, angled with the ovary .... *Enc. duveenii*
65. Not as above species .... go to 66

TABLE IV *(continued)*

66. Plants smallish with 2–3 leaves, flower stalks only a little higher; flowers distinguished by long, narrow, bowed lateral lobes with scalloped tips and high fragrance of violets; midlobe from a long isthmus and with a central callus ridge, fanlike to circular shape with an indented apex, and serrate margin; sepals and petals green-brown with a yellow margin . . . . *Enc. ionosma*
66. There are still other choices . . . . go to 67
67. Plants shade-loving with flowers that turn black on drying; terrestrial species with 2 leaves of unequal length reaching about 18 cm high; sepals and petals distinctly obovate; long, oblong lateral lobes with scalloped tips; midlobe rounded from a tapered base forming the isthmus with its callus ridges; apex slightly indented and midlobe with irregular edge; column curved, with auricles, 12 mm long . . . . *Enc. latipetala*
67. None of the above . . . . go to 68
68. Tall panicles can reach 1 m with growths having 2 leaves; petals about half the width of the dorsal sepal, both oblanceolate-obovate; prominent callus ridges extend over the isthmus about halfway down the disc of the midlobe, continuing as the 3 main veins; midlobe of lip round, apiculate and wavy-edged; lateral lobes oblong-ligulate with rounded tips that are a little dilated; bold purple veining on the white midlobe . . . . *Enc. maderoi*
68. A few more combinations . . . . go to 69
69. Plant small with panicles 1.5 m long; sepals and petals brown with green bases and a narrow rim; lip with some green at the base, purple veining, and a wide white margin; oblanceolate dorsal sepal about half the width of the obovate petals, broad-shouldered triangular rounded lateral lobes, each almost as large as the midlobe; midlobe concave around the midvein; column white with brown-rose flushes toward the base . . . . *Enc. meneziana*
69. Otherwise . . . . go to 70
70. Large plants with 2–3 leaves and fleshy flowers in tall panicles; sepals and petals green-brown with a yellow edge; lip white with a few purple striations; dorsal sepal obovate, petals spatulate and about half is narrow clawed base; lateral lobes of lip falcate-oblique, veined at the base, white at their tips; midlobe circular, notched at the apex, somewhat crenate; column 10 mm long, with prominent auricles and a bright yellow-orange anther cap . . . . *Enc. naranjapatensis*
70. Still a few combinations to investigate . . . . go to 71
71. Large plants with tall panicles and fleshy flowers; leaves without an abscission zone so must be cut to remove from plant; flowers turn black on drying; sepals and petals olive green with brownish veining on some clones; petals obovate and clawed; lips white aging to cream and with 3 short purple lines beneath the callus; midlobe rounded to oval, notched at the tip; callus ridges extend to the apex; ovary surface smooth; flowers strongly fragrant . . . . *Enc. pachyantha*
71. Not as above . . . . go to 72
72. Plants smallish for flower size, with 2 leaves; 4–5 flowers on a short raceme; narrow sepals and petals green-brown, can be reflexed, curving backward from the lip; lip white to yellowish to slate blue with purple veining; midlobe wider across at the apex than the tapered base and with a distinct notch where the central vein from between the 2 callus ridges ends; veining on the disc granular- verrucose; column whitish, 8 mm long, with a white anther cap . . . . *Enc. pauciflora*
72. We are on the home stretch . . . . go to 73

TABLE IV *(continued)*

73. Flowers thin-textured, almost translucent, lime green with white lip, borne in a panicle; ligulate lateral lobes narrow with rounded tips; midlobe round, indented at the tip, and purple veined; column straight, with auricles; anther cap bilobed, with maroon markings that can look like eyes .... *Enc. profusa*
73. Not listed above .... go to 74
74. Plants medium-sized with panicles of fleshy flowers to 50–60 cm or more; petals wider than the dorsal sepal and more spatulate-obovate in shape; lip with a short claw; sepals and petals yellow-green, lip white with purple veining; lateral lobes narrowly ligulate, oblique with rounded tips; midlobe tending to a transverse oval shape, notched at the apex .... *Enc. steinbachii*
74. Not as above .... go to 75
75. Pretty, large, fragrant flowers; sepals and petals cinnamon-brown; lip rosy pink with darker veining; tips of the lateral lobes expanded with scalloped ends; midlobe broadly rounded with the base wider and a deep notch at the apex, veining concentrated into a central large spot on the disc; petals obovate with short clawlike base .... *Enc. spiritusanctensis*
75. Plants and flowers otherwise .... go to 76
76. Flowers on a tall raceme to 70 cm with 8–10 distant flowers; sepals and petals greenish red-brown as are the lip lateral lobes; midlobe rosy magenta, oval, a little longer than wide; central veins thickened and continuing to apex of midlobe; petals narrower than sepals; column a little humped, with small auricles and a rose-pink anther cap that is deeply divided .... *Enc. seidelii*
76. Your last chance to choose .... go to 77
77. Sepals and petals green-brown but have curly edges and wide green bases about one-third their length; lateral lobes large, wide, and well shouldered with nearly truncate tips, green below, with yellowish veins in their upper half; midlobe wavy edged, rounded, and notched at the apex; column green basally with rosy flushes toward the tip, wingy edged with a small projection or tooth .... *Enc. schmidtii*
77. Short plants with one, sometimes 2 leaves and short racemes with only a few flowers; sepals and petals greenish gold, lip white with rose streaks; petals spatulate, about twice the width of the dorsal sepal; lip midlobe tending to an elongated oval with raised veins, especially the central 3; column about 18–19 mm long with auricles; ovary surface smooth; flower parts turning brown, not black, on drying .... *Enc. remotiflora*

CHAPTER 2

Description of Species

Index of South American Species of *Encyclia*

Acute Encyclia

Brazil, Venezuela

(FIGURES 1-1, 2-1)

Encyclia acuta Schlechter. 1925. *Beih. Bot. Centralbl.* 42(2):104.

Synonym

Epidendrum acutum (Schlechter) Hawkes. 1956. *Orquídea* 18:174.

We know apparently little about this species, and except for the type description and the account by Carnevali and Ramírez (1988), literature on the species is meager. It is not described by Cogniaux in *Martius*, or by Dunsterville and Garay in *Venezuelan Orchids*, or by Foldats in *Flora de Venezuela.* Carnevali and Ramírez tell us that plants are relatively frequent in the eastern region of Bolívar state and the delta of the Orinoco River. They grow as epiphytes at low to moderate elevations in woods with shaded exposure to the sun. The species is least common in the Amazonas area where it has been found only a few times. The type specimen, G. Huebner #49, was collected near Manaus, Brazil, flowering in June. It is a small species: plants with flowers may be only 5 cm tall with their flowers—or 10 cm tall in a larger plant. The leaves are relatively leathery, and the panicles are above the leaves with the stalks angular in section instead of the usual round. The flowers are proportionately small.

The sepals and petals are green, sometimes with brown stains, and the lip is greenish white with one to three lilac-purple veins. The midlobe is suborbicular and apiculate. Auricles are present on the column. The drawing in Carnevali and Ramírez is the only illustration. Though Carnevali and Ramírez have some reservations about their specimens matching the *acuta* type, now destroyed in Berlin, they feel it is a likely match even though they cannot be certain. A neotype specimen needs to be designated. Castro (1998) has written that *acuta* is a synonym of *Encyclia chloroleuca.* Comparison of their lip outlines certainly shows a close similarity, and further study of the types could confirm this proposal. Castro also includes *Enc. viridiflava* as another synonym (please see).

The type description gives the following measurements and shapes: sepals oblong-ligulate, acute, seven-nerved, 14 mm; lip about 14 mm long; petals oblique oblong-spatulate, seven-nerved, 14 mm; lateral lobes of lip

Figure 2-1. *Encyclia acuta.* Habit sketch and flower detail. With permission of Germán Carnevali.

12 mm long and 14 mm between tips when lip is flattened; midlobe about 8 mm long × 5 mm wide; column 6 mm long; pedicellate ovary 20 mm long. The main vein of the callus goes to the apex of the lip, and the sides of the midlobe are reflexed.

The description might well fit one of the other species, and we may never know for certain if these plants are properly named. In Hoehne's (1952) list the species is described as being from the Amazonas area and regions of northern Brazil. Schlechter also cites Huebner #153 from Manaus and Huebner #128 from Iguapo Caicara. Schlechter wrote that *Encyclia acuta* is closest to *Enc. longifolia,* which has much larger, stronger plants and larger flowers and is currently called *Enc. oncidioides.* He also states that *Enc. acuta* is best recognized by its small flowers, short branches, short flower stalk, and pointed sepals, petals, and leaves. Nothing else is written about this species that I have been able to find, so it really is on the borderline between the known species and the dubious species in Chapter 3.

Advent Encyclia

Brazil

(FIGURE 1-1; PLATE 1)

Encyclia advena (Reichenbach f.) Porto and Brade. 1935. *Rodriguésia* 1(2): 28.

Synonym

Epidendrum advenum Reichenbach f. 1872. *Gard. Chron.,* p. 1194.

This species is one of the large-flowered encyclias of Brazil that have been confused with one another over the years. The other species include *Encyclia osmantha, Enc. megalantha, Enc. pachyantha, Enc. randiana,* and perhaps *Enc. albopurpurea.* The hallmarks for distinguishing this species are the large orbicular to obovate-suborbicular petals with a short claw, the obovate broad sepals, the narrow wasp-waist isthmus, the entire rounded tips of the lateral lobes wider than the basal portions of the lobes, and the broad transversely oval midlobe with a retuse or notched apex with no apicule and a somewhat ruffled edge. The column is humped in shape and has distinct auricles that are white. The ovary surface is smooth.

The pseudobulbs and foliage were not described by Reichenbach as he was apparently sent only the flowers by John Day. The panicles are long and branched, three or four times the length of the leaves. The original plant flowered with O. O. Wrigley at Bridge Hall in Bury, Lancashire, grown by a Mr. Hubberstey, and sent on to Reichenbach for identification. The type specimen is conserved in Vienna with two other examples on the herbarium page, each with outline sketches and notes by Reichenbach. It must have flowered first during Advent, the period including the first four Sundays before Christmas, accounting for its name.

My notes show the flowers fleshy, cupped, and with the sepals and petals olive green to yellow with brownish maroon veining and flushes. The lip is white with purple veining on the midlobe and over the green bases of the lateral lobes. The column is green with maroon flushes and a white tip. The anther cap is yellow. Measurements of the flower parts show variability among clones. One of the largest flowers I have seen had the following measurements: sepals 22 × 9 mm; petals 23 × 15 mm; lip 26 mm long including a 3 mm claw; lateral lobes 11 mm long × 5 mm across their bases and 9 mm across the apices; midlobe 16 mm wide × 13 mm long; and the isthmus 3 mm. The column is 10 mm and the pedicellate ovary is 34 mm. Some clones have almost spade-shaped petals, more pointed than the others toward their apices.

John Day's 1883 painting (Volume 37, t. 35) of the original plant of *Encyclia advena* from Low and Company shows the sepals and petals yellowish tan with full tessellation of radiating brown veins. The flowers are unusually large, and the lip is yellowish white with purple veining in the midline. The column is green. Rolfe said that the flowers of *Enc. advena* were most similar to those of *Enc. osmantha.*

I cannot explain why, but Fowlie and Duveen (1992) have made *Encyclia advena* a synonym under *Enc. megalantha,* a large-flowered but, I think, completely different species. I am considering them separate species in this volume and the *advena-osmantha* conflict will remain for future resolution.

White and Purple Encyclia

Brazil

(FIGURE 1-1; PLATE 2)

Encyclia albopurpurea (B. Rodrigues) Porto and Brade. 1935. *Rodriguésia* 1(2):28.

Synonym

Epidendrum albopurpureum B. Rodrigues. 1877. *Gen. & Spec. Orch. Nov.* 1:51.

As I have mentioned elsewhere, João Barbosa Rodrigues early on published the names of many new orchid species as the result of his local travels, and he knew much about the orchids of Brazil and was an early director of the botanical garden in Rio de Janeiro. He produced six volumes of Brazilian orchid paintings from 1868 to 1873, but they were only published recently (Sprunger et al., 1996). One of the original volumes, dealing mostly with the Laeliinae, is conserved in the rare book library at Harvard University, having been purchased by Oakes Ames for his orchid library. The other volumes are now at the botanical garden in Rio de Janeiro. The text covering the plates, however, was not published by Rodrigues until 1877 for Volume I and 1882 for Volume II, the latter dates thus becoming the valid dates of publication for his species.

When Alfred Cogniaux was preparing his monumental coverage of the Brazilian orchid species for Martius's *Flora Brasiliensis,* published from 1898 to 1902, Rodrigues sent his plates to Cogniaux, who in turn used them to illustrate the Martius orchid volumes that were parts 4, 5, and 6 of Volume III of that work. The plates were sent back to Brazil via the orchid herbarium at Kew. The paintings were copied faithfully by Harriet Dyer, the wife of Sir William Thistleton Dyer, director of Kew at the time, and have been a treasured source for our reference records ever since. Only one plate has been found missing of all the species of *Epidendrum* that Rodrigues described—*Epi. albopurpureum*—and no one knows when or where it disappeared, or, actually, if it ever existed, though Rodrigues cites t. 263 in his type description. We do have, however, the description to give us the plant and flower details. A fire at the botanical garden in Rio de Janeiro may have destroyed the type specimen, as it is also missing.

Rodrigues states that the plants came from the river Tapajós at Uricurytyba in Pará state, growing on calabash trees and flowering in September. He described the plants with oval pseudobulbs, two or three linear, acute leaves, and lanceolate petals with margins undulate and recurved tips. The lips had oblong lateral lobes with pointed tips, and the midlobe was subreniform, undulate, and notched at the tip. The flower panicles were three times longer than the leaves and emitted a strong fragrance of vanilla. The sepals and petals were yellow-brown, and the large lip was white with much carmine veining. The column was winged at its apex. It was, altogether, a beautiful species. Cogniaux in *Martius* repeats what Rodrigues has told us.

More recently, Jack Fowlie and Denis Duveen (1992), in their research on the Brazilian encyclias, have attempted to identify this species, mostly by having eliminated all the known ones from the Brazilian taxa seen in various collections. Luckily, they found a plant belonging to Jorge Verboonen at Orquidário Binot with almost rounded, pyriform pseudobulbs and large flowers on a tall scape, flowering in September, the Brazilian spring. It had been brought from the river Tocantins, a location also in Pará state. It is illustrated in color in the *Orchid Digest* (Oct.–Dec., 1992) and clearly fits the description of *Encyclia albopurpurea.* The sepals and petals are obovate-lanceolate, acute, and a rich red-brown. The lip is white and nearly covered with carmine veining and flushes; the veins at the base of the lateral lobes and the five central veins of the midlobe show darker maroon markings; the lateral lobes are oblong with acute apices. The lip is reflexed along the central vein of the disc. The column appears white with a rosy flush, and the auricles are pale.

I hope that Orquidário Binot will raise selfed seedlings of this rare species, which appears to be well named and well described. We must find more herbarium material for the botanical records and establish a neotype record for the species. A publication would now be in order to confirm this identification and to designate a specific specimen as the neotype. While we are so involved with the fieldwork, perhaps an observation can be made on the pollinators that prefer a saddlelike lip with the sides of the midlobe curling down. That characteristic is common among the larger-flowered encyclias and probably provides some larger bee a better foothold or "saddle" from which nectar can be obtained and pollination, incidentally, takes place. Few field observations on pollinators are available for consultation. We also need measurements of the flower parts.

Two-horned Encyclia

Argentina?, Bolivia, Brazil, Paraguay, Peru, Venezuela

(FIGURE 1-1; PLATE 3)

Encyclia amicta (Linden and Reichenbach f.) Schlechter. 1919. *Repert. Spec. Nov. Regni Veg.*, Beihefte 6:75.

Synonyms

Epidendrum amictum Linden and Reichenbach f. 1855. *Bonplandia* 3:219.
Epidendrum linearfolioides Kraenzlin. 1911. *Kungl. Sv. Vet. Akad. Handlingar* 46(10):55, t. 9, fig. 4.
Encyclia linearfolioides (Kraenzlin) Hoehne. 1938. *Arq. Bot. Est. São Paulo* 1(1):19.
Epidendrum amicta Carnevali and Ramírez subsp. *amicta*. 1988. *XLVI Exposicíon de Orquídeas, Comité de Orquideología,* Bol. No. 23:13–87.

A few *Encyclia* species are quick and simple to recognize because of a single hallmark characteristic, namely, a very deep maroon-colored anther cap that appears nearly black. When a species has this quality, it is usually noted, especially if the cap has two rather protruding "horns" or extensions with a groove between instead of the usual gently rounded halves that cover the two cells of the anther. As I delved among the South American species with these two characteristics—the anther cap color and the protuberances—I ran into one problem after another. I ended up with a large group of synonyms for a species that was first named *Epidendrum amictum* in 1855, predating all the other names and having therefore the priority of epithet for this species. Not all may agree with this choice, and some, including myself, will choose to keep some of the species of this complex as separate entities. Such species are described on other pages under *Encyclia conchaechila, Enc. leucantha,* and *Enc. thrombodes.* If we consider all these closely related forms together, the complex has a major distribution in South America, though local populations may differ in some details. *Encyclia linearfolioides* continues to be a debatable synonym, and Fowlie and Duveen (1992) describe it as having an anther cap with spots that can look like eyes, a minutely notched lip, and white flowers 2 cm in diameter or less.

We can also wonder what might be different about the pollinator of these plants. Ordinarily the anther cap of an *Encyclia* species is white, greenish, or sometimes yellow, and dark colored types such as these deep maroon caps are unusual. Since the base of the lip tends to be covered with short, dense, almost microscopic hairs, especially on the forcipate keels, and since the midlobe is characterized as having distinctly raised and verrucose veins, the lip is also distinctive. Furthermore, the lateral lobes tend to be somewhat elongated and clasp tightly about the column. The lateral lobe shape and attitude, and the geographic gap separate, for me, *Enc. conchaechila, Enc. leucantha,* and *Enc. thrombodes* from *Enc. amicta.* Castro (1992) considers *Enc. bicornuta, Enc. microxanthina,* and *Enc. argentinensis* as synonyms, an idea that does not match our observations of these species.

As indicated for other species, some fieldwork on natural pollinators is desirable. We need to know more about the significance of these often small variations and what answers they may provide about species differences. Past experience has shown, usually, that any part of a flower that is directly involved in pollination—and therefore the perpetuation of the species—remains unchanged or unmodified from plant to plant, and flower to flower.

The flowers of *Encyclia amicta* have green to tannish sepals and petals, sometimes with a pinkish flush, and they are narrowly lanceolate in shape with pointed tips. In my examples they measure about 2 mm wide and 15–16 mm long. The lip is about 13–15 mm long with a rounded and pointed midlobe and lateral lobes that flare apart when the lip is flattened for study. Well-grown specimens may have slightly larger flowers. Usually the margins of the lateral lobes have a slight indentation or break about 1–2 mm from their apices. Also, a distinct isthmus is noted between the base of the lip with its callus and tiny hairs and the beginning of the midlobe. The lip is a creamy white, tending to yellow as it ages, and the keeled central veins are slightly green to purple, sometimes on a pink-flushed background. The somewhat verrucose and keeled central vein emerges from between the two side portions of the callus and extends clearly to the pointed apex. Two or three lateral veins on either side of it help support the central regions of the midlobe. The column is slightly curved, has no auricles, but does have slightly winged edges. The leaves are to 24 cm long × 6–7 mm wide, and the slightly nodding racemes or panicles may be somewhat higher than the leaves.

The plants tend to be summer bloomers and, as with others of this genus, tend to grow best when mounted in a basket or on a slab of cork or tree fern. A specimen plant of Wynn McPheeters was granted an 83 point CHM by the American Orchid Society in 1991. It had 185 flowers and 20 buds on eight flower stalks and was grown in a 10 cm basket in St. Petersburg, Florida. The type specimen was cultivated by Linden, and the Reichenbach Herbarium #936 specimen from the herbarium of the Museum Palatinum Vindobonensis (Royal Museum in Vienna) is marked Linden and Reichenbach with a drawing.

Andrich's Encyclia

Brazil

(FIGURE 1-1; PLATE 4)

Encyclia andrichii Menezes. 1992. *Orchid Digest* 56(3):148.

The holotype specimen of this species was collected at about 200 m in São João de Petrópolis in Espírito Santo state, Brazil, and it flowered in Menezes's lath-house in October, 1991. It is named after Aldo Andrich from Vitoria City who found the first plant. The type specimen is conserved at the herbarium of the University of Brasilia.

The plants are epiphytic and have two or three erect leaves 25 cm long × 2 cm wide with acute tips. The flower stalk can be 60–100 cm tall with a panicle of 15–20 slightly fragrant flowers. The sepals and petals are greenish brown and darker toward the pointed apices, the midlobe white with rosy tints, ruffle-edged, and veined with red-purple. The lateral lobes are yellow-green marked with brown veining. The colored veining covers both the midlobe and the lateral lobes except for a narrow white margin.

The sepals are obovate-oblanceolate, 15 mm long × 5 mm wide. The petals are subspatulate, 14 × 4 mm. The midlobe is rounded with verrucose veining, and the callus is also marked with spots of color. The column is 8 mm long and greenish white. The type description does not mention lip length, any auricles on the column, or the surface of the ovary.

No other details are available, and apparently the plants have not been in general cultivation. Looking at the flattened lip, I am reminded of the flowers of the Mesoamerican species *Encyclia parviflora.*

Narrow-lobed Encyclia

Bolivia, Ecuador, Peru

(FIGURE 1-1; PLATE 5)

Encyclia angustiloba Schlechter. 1921. *Repert. Spec. Nov. Regni Veg.,* Beihefte 8:89. Also see Mansfeld, 1929, *Atlas der Kordillerenstaaten,* t. 94, no. 369.

Synonyms

Encyclia mapiriensis Kraenzlin. 1928. *Repert. Spec. Nov. Regni Veg.* 25:21.
Epidendrum strobelii Hawkes. 1956. *Orquídea* 18:173.

Misapplications (*auctor non*)

Epidendrum angustilobum Fawcett and Rendle. 1909.
Epidendrum guentherianum Kraenzlin. 1928. *Repert. Spec. Nov. Regni Veg.* 25:20 = *Encyclia mapiriensis* = *Encyclia angustiloba.*

Dodson and Dodson (1980) describe this *Encyclia* species as an uncommon epiphyte in dry thorn forest on cactus in western Ecuador. It is almost unknown in herbaria, but we have the type description with a drawing in the *Atlas,* and another drawing in the Dodson *Icones.* The leaves can be red in the bright light, and there are one or two leaves 25 cm long × 1 cm wide. The bracts are scarious and weather into stringy fibers. The panicle is gracefully upright and has short densely flowered branches.

The flowers are on the small side for this genus, the sepals and petals red-brown, and the lip white. The longish slightly curved column has auricles, and the ovary surface is comparatively smooth except for a few papillose spots nearest the sepals. Measurements of a specimen observed at F. L. Stevenson's greenhouse are as follows: dorsal sepal 16 × 5 mm, subspatulate-obovate; petals 15 × 6 mm, more spatulate and clawed than sepals; lip slightly clawed, 15 mm long. The midlobe measures 6 mm broad and about 5 mm long emerging from a 3 mm isthmus.

The long lateral lobes are immediately distinguishing with their parallel sides and obtuse apices that are somewhat scalloped, asperous, and rough or crenate, 8–9 mm long × 1.5–2.0 mm wide. When the lip is flattened, the lateral lobes extend 15 mm from tip to tip, extending to a level almost halfway down the midlobe. The midvein between the forcipate main callus ridges runs to the obtuse apex of the rhombic-oval midlobe. Two major veins are on either side of the raised central one. The column is 10 mm, and the roughly asperous pedicellate ovary is 20 mm long.

This species appears to have similarities with *Encyclia ionosma* (please see) especially in the lip configuration. The latter comes from northern Brazil and the Guianas while *Enc. angustiloba* comes from the Andean regions of Bolivia, Ecuador, and Peru. The Ecuadorian plants grow in dry scrubby desert environment. How do I tell them apart without knowing about their geography?

Kraenzlin's *Epidendrum guentherianum,* collected around Mapiri, San Carlos, near Larampinni at 600 m in Bolivia, is, presumably, in the type description, based on Buchtien #490, but the actual specimen is misnumbered as #496. *Epidendrum guentherianum* was described as a reedstem epi and cannot ever be considered as an *Encyclia.* Two herbarium specimens are involved in the mixup, one at Kew and one in Vienna. The #490 is an *Encyclia* I had tentatively identified as *Enc. mapiriensis,* now a synonym of *Enc. angustiloba.*

Argentine Encyclia

Argentina, Brazil, Paraguay

(FIGURES 1-1, 2-2; PLATE 6)

Encyclia argentinensis (Spegazzini) Hoehne. 1952. *Arq. Bot. Est. São Paulo* 2(6):151.

Synonyms

Epidendrum argentinense Spegazzini. 1916. *Ann. Mus. Hist. Nat. Buenos Aires,* p. 125, t. 28.

Encyclia saltensis Hoehne. 1938. *Arq. Bot. Est. São Paulo* 1(1):19, t. 10.

Misidentification (*auctor non*)

Epidendrum oncidioides var. *gravidum* = *Epidendrum gravidum* = *Encyclia gravida* (see Vol. IV, p. 18).

This strong-growing species has pyriform pseudobulbs and long narrow leaves. Sometimes the pseudobulbs are spaced apart, not in a tight cluster. The panicles may have one or two short branches and are about the length of the leaves or a little longer. The flowers are comparatively large and easily recognized, for the midlobe of the lip is flat, longer than wide, and has a slightly ruffled border. The column is broad and heavy and has auricles. The callus on the lip is basinlike between the two outer ridges, and three main veins emerge from the three teeth at its tip and run to the blunt apex. Two additional veins may be noted as enlarged, but they emerge from the side of the callus ridges and then spread out on the midlobe. The specimen named *saltensis* was collected in Salta, Argentina, and apparently the species was given two names before it was realized that they were synonymous. The early confusion with *Encyclia oncidioides* was just a case of mistaken identity that found its way into publication.

The sepals and petals are a dull greenish yellow with obscure maroon veinings, and the lip midlobe is white with violet or purple markings over the major veins. The base of the lip is greenish, and the lateral lobes are veined about halfway to the white or cream-colored tips. The column is pale with some reddish or purple marking, is clearly auricled, and the ovary surface is rough (asperous). The flower stalks are reddish and roughened with white-tipped papillae.

Some measurements and comments taken from the very complete and detailed type description are as follows: pseudobulbs conical, 30–40 mm tall × 1.2–2.5 mm wide; leaves usually two, narrowing toward the roundish point, 30–40 cm long × 0.6–1.0 cm wide; scape about 40–45 cm tall with two to three short branches. Sepals are 6.0–6.5 × 2 mm, oblanceolate and pointed, five-veined; petals oblanceolate, pointed, five-veined, and a little shorter than the sepals, 23 × 6 mm; lip 20 mm long with 3 mm claw, 1.5 mm isthmus, and midlobe elliptic-subquadrangular, 9 mm long × 10 mm broad, with a slightly pointed apex and an undulated and slightly toothed border; lateral lobes of lip are strap-shaped with rounded tips; the column is about 10 mm and the ovary and pedicel about 30 mm.

According to the herbarium labels, this species is found at low elevations. Although nothing is said about the kind of habitat it prefers, I would assume warm, bright situations. The type specimen was collected from the thick branches of *Apuleia* trees. Little else seems to be known about it.

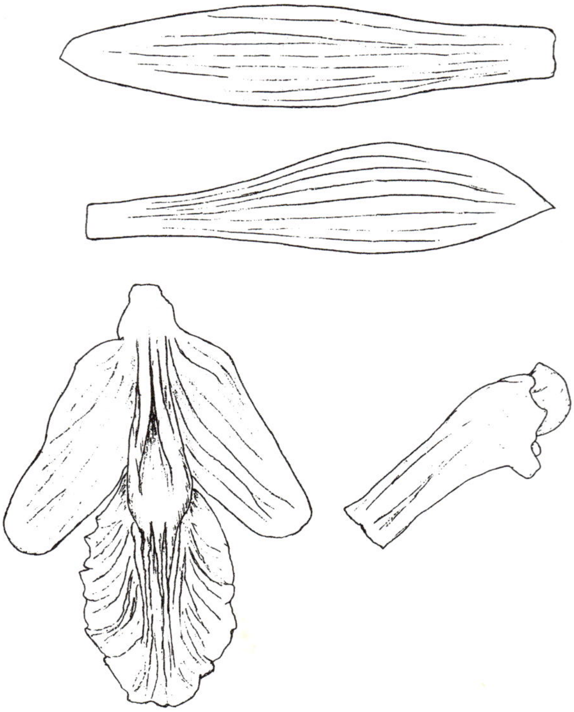

Figure 2-2. *Encyclia argentinensis*. A flower from the type specimen of *Enc. saltensis*, a synonym, conserved in the Lindley Herbarium at Kew Gardens. *Top:* dorsal sepal. *Middle:* petal. *Bottom, from left to right:* lip and side view of column. Camera lucida drawing by Antonio Toscano do Brito.

Rough-stalked Encyclia

Colombia, Ecuador, Peru

(FIGURE 1-1; PLATE 7)

Encyclia aspera (Lindley) Schlechter. 1918. *Beih. Bot. Centralbl.* 36(2):471.

Synonyms

Epidendrum asperum Lindley. 1834. *Hooker Journ. Bot.* 1:6.
Encyclia trachypus Schlechter. 1916. *Repert. Spec. Nov. Regni Veg.* 14:393.
Also see Mansfeld, 1929, *Atlas der Kordillerenstaaten*, t. 47, fig. 185.
Encyclia asperirachis Garay. 1953. *Sv. Bot. Tidskr.* 47(2):220, fig. 10.
Epidendrum asperirachis (Garay) Hawkes. 1956. *Orquídea* 18:174.

In spite of the synonyms, the identification of this species over the years has been little questioned. Its Andean location, the distinctly rough and reddish flower stalk, and the well-spaced brownish ocher to dull yellow-colored flowers widely spaced in panicles, or racemes, with their raspberry or cherry fruity fragrance, can be easily discerned. The roughness or verrucosities present on the flower stalks are also apparent in the raised veins that are on the midlobe of the lip and on the ovaries. Also, there are not so many Andean *Encyclia* species with which it can be confused. The type specimen #1250 was collected by Hugh Cuming in western Colombia in 1831, and it is in the Lindley Herbarium at Kew Gardens. The species's existence in Panama has been debated, but I have not yet found any Panamanian specimens in any herbaria. It does not seem likely.

The plants have two narrow leaves to 35 cm long × 1.5 cm wide, and the rough-surfaced panicles with zigzag fractiflex branches extend to twice or three times the leaf length. The branches have five to seven flowers. A boiled-up flower for dissection shows that the sepals are lanceolate to oblanceolate in shape, about 5.5–6.0 × 16 mm, with pointed tips. The petals are more clawed and spatulate, measuring about 4.5–5.0 mm across the wider tips and 1 mm at their base and 16 mm long. The flattened lip is 16 mm long, has a slight claw, lateral lobes spreading to 17 mm, and a rounded midlobe 8–9 mm long × 9 mm wide. The tips of the lateral lobes turn up a little, are clear of veining, and are rounded but pointed in shape as their edges turn back. The lip can be whitish or with a yellow and green background, and the five to seven central veins tend to be faintly colored with red-purple. The short, straight column is about 5–6 mm long, winged, and has yellow auricles. The anther cap can be marked with small spots of maroon or red, but is not the so-called black or horned cap that is characteristic of certain other species (see *Encyclia thrombodes*, for example).

The plants seem to be comparatively common around coastal central Ecuador and do well mounted on cork. They can dry out, but prefer a humid atmosphere. The clone 'Lil' was granted a CBR by the American Orchid Society and had 50 flowers and three buds on one panicle. The flowers had a natural spread of 2.6 cm.

Auyán-tepui Encyclia

Venezuela

(FIGURES 1-1, 2-3)

Encyclia auyantepuiensis Carnevali and Ramírez. 1994. *Lindleyana* 9(1):61.

This species is named after the Cerro Auyán-tepui, one of the flat-topped mountain tepuis in Venezuela and the only known location for this species. It grows mostly on rocks, but may also occasionally be found as an epiphyte on trees, and it enjoys bright sun. The authors describe the plants as having short rhizomes noticeable between growths, two leaves, and either racemes or panicles with thin, erect stalks and well-spaced widely starry flowers. Drawing #456 by Dunsterville portrays it well, although the plate is identified in the Dunsterville and Garay *Field Manual* as *Epidendrum oncidioides.* It is in the regular Volume II on page 151. Carnevali and Ramírez tell us that of the two flowers on the original plate, the right-hand flower is this species and the left-hand flower is something else. I would agree with that, and I also believe that neither flower represents *Encyclia oncidioides.*

Dunsterville's drawing #456 also shows the column, widened at the apex and bearing a "black" anther cap that is horned, but the column is not auricled or winged. No mention is made of hairs or pubescence on the lip callus, and the type description says it is smooth; also the ovary surface is smooth. The keels of the callus lead to three main veins that extend toward the tip over the disc of the midlobe. The lateral lobes are large in proportion and taper to a blunt point with the colored veining extending about halfway down the lobes. The sepals and petals are purple-brown in color with a light yellowish border and with occasional scattered fine white hairs on their inner surfaces. The dorsal sepal measures 15 × 4 mm, the petals 13.5 × 3.5 mm, and the lip is 12 mm long. The distance between tips of the lateral lobes is 13 mm, and the midlobe measures around 16 mm long × 8 mm wide. The base of the lip is greenish with the purple veining, and the midlobe is white with the raised veins also purple.

The species is not widely known nor grown and, like so many of these South American species, needs more attention. It appears close to *Encyclia amicta.*

Bracted Encyclia

Brazil

(FIGURES 1-1, 2-4; PLATE 8)

Encyclia bracteata Schlechter *ex* Hoehne. 1952. *Arq. Bot. Est. São Paulo* 2(6): 150.

Synonyms

Epidendrum bracteatum B. Rodrigues. 1877. *Gen. & Spec. Orch. Nov.* 1:52.
Epidendrum pusillum Rolfe. 1891. *Gard. Chron.*, ser. 3, 10:669.
Epidendrum pabstii Hawkes. 1956. *Orquídea* 18:172.
Encyclia hoehnei (Hawkes) Pabst. 1975. *Bradea* 2(6):21.

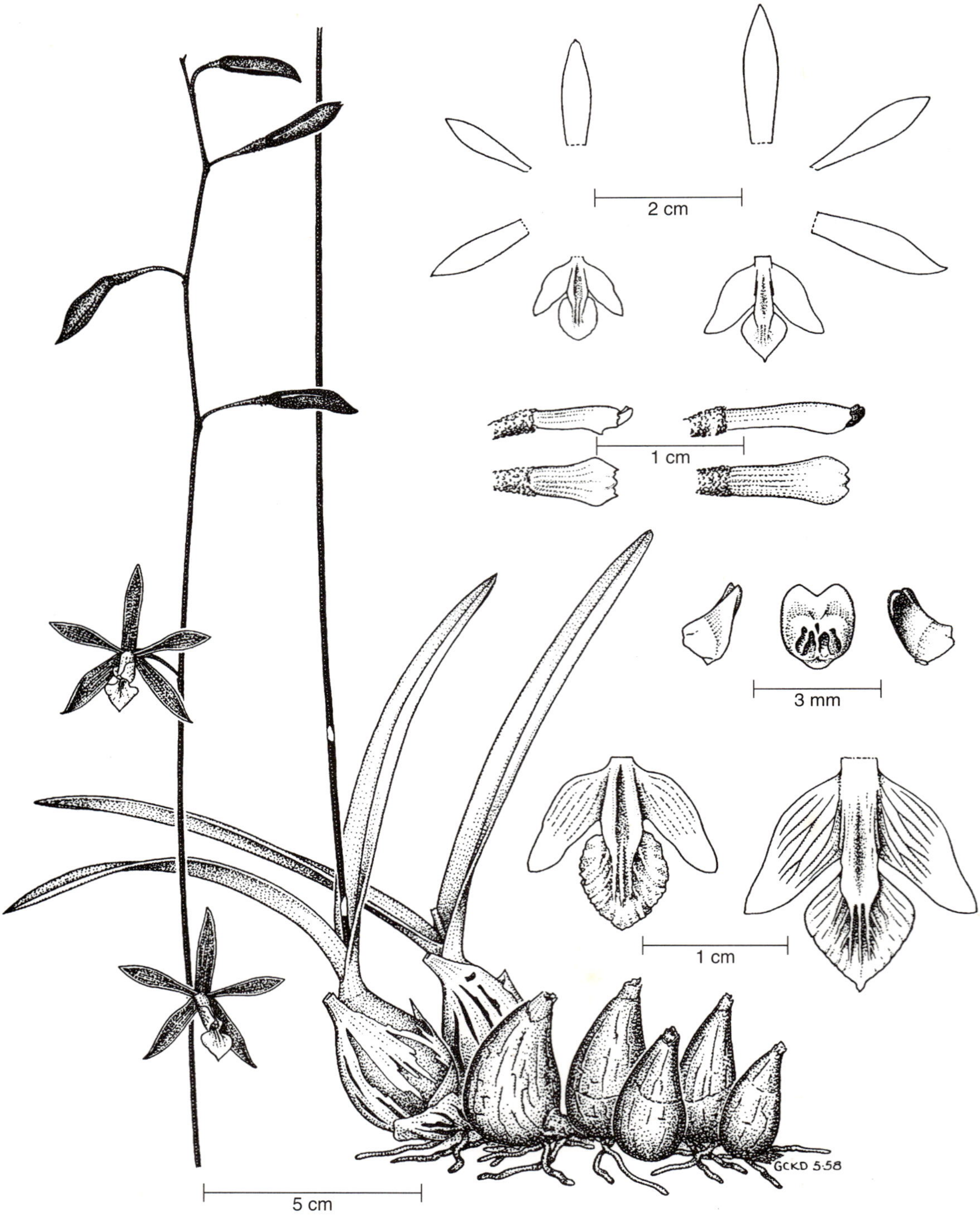

Figure 2-3. *Encyclia auyantepuiensis*. Drawing #456 by G. C. K. Dunsterville.

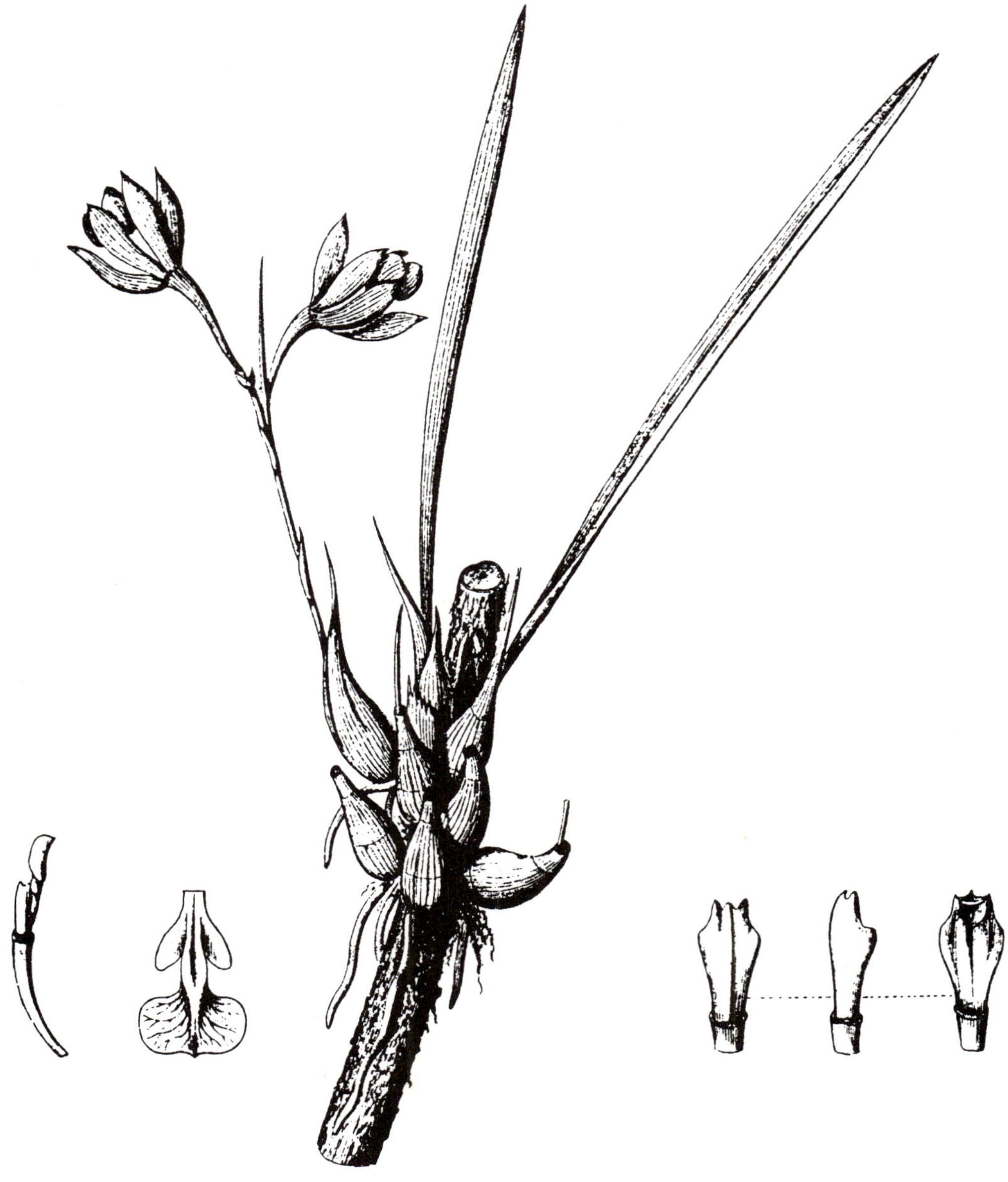

Figure 2-4. *Encyclia bracteata.* Type drawing by Barbosa Rodrigues published as *Epidendrum bracteatum* by Cogniaux. Adapted from Plate 14 of *Martius* 3(5).

Misapplications (*auctor non*)

Epidendrum bracteatum Vellozo. (1825) 1827. *Fl. Flum.* = *Rodriguezia venusta* Reichenbach f. (see Cogniaux in *Martius* 3:153).

Epidendrum pusillum Linnaeus. 1763. *Species Plantarum* = *Psygmorchis pusilla* (Linnaeus) Dodson and Dressler.

This miniature and unique species is one of the smallest vegetatively in the genus, with its lax, narrow, almost terete leaves from thin bases and thin flower stalks usually with only a single flower. It has an undeniable charm. The green-brown mottled sepals and petals with greenish rims contrast clearly with the bright rosy magenta on the lip. I have never understood why the flowers have not been used more to make miniature hybrids. The plant is small, brightly colored, and easily grown on a slab of cork or in a basket. Flower number can increase to two or three on stalks of well-grown plants.

Although readily recognized and often grown, *Encyclia bracteata* has had little written about it since Barbosa Rodrigues described it as "insignificant but to be recommended because of its flowers." The original plants came from Espírito Santo state and flowered in October. Cultivated plants can flower in late summer through early fall. The small lateral lobes of the lip are white with green borders, and the callus is also white, the whiteness extending along the central vein to the apex of the lip. The midlobe starts from a wedgelike base expanding to a broadly oval shape. The 8–9 mm column is broader near the tip and winged, the wings producing auricle-like protrusions and then terminating in more pointed extensions beside the anther cap. Sepals and petals are 15–16 mm long, the oval of the midlobe being about 9 mm wide × 7 mm long.

Bradford's Encyclia

Tobago, Trinidad, Venezuela?

(FIGURES 1-1, 2-5)

Encyclia bradfordii (Grisebach) Carnevali and Ramírez. 1986. *Ernestia* 36:9.

Synonym

Epidendrum bradfordii Grisebach. 1864. *Fl. Brit. W. Indies,* p. 613.

Although this species was described in Volume IV from Trinidad with a questionable occurrence in Venezuela, it has now been reported by Carnevali and Ramírez (1988) from the coastal mountains. It is a strange species in that some of the flowers may open partially, but then the others on the raceme or panicle immediately begin to develop cleistogamously. One expects that the other fat green buds will continue to open, but in my greenhouse they do not. All the ovaries begin to enlarge, and the buds stop growing and do not open—a disappointment for all. The capsules are somewhat twisted, plumply rounded, bright green, and smooth to the touch. The small flowers have bright apple green sepals and petals, and the flattened lip is slightly more white with its little pointy lateral lobes extending outward and down. The sepals and petals only measure 12–13 mm long, and the lip is 13–14 mm long. There are large auricles on the column. See Volume IV for further discussion about the species.

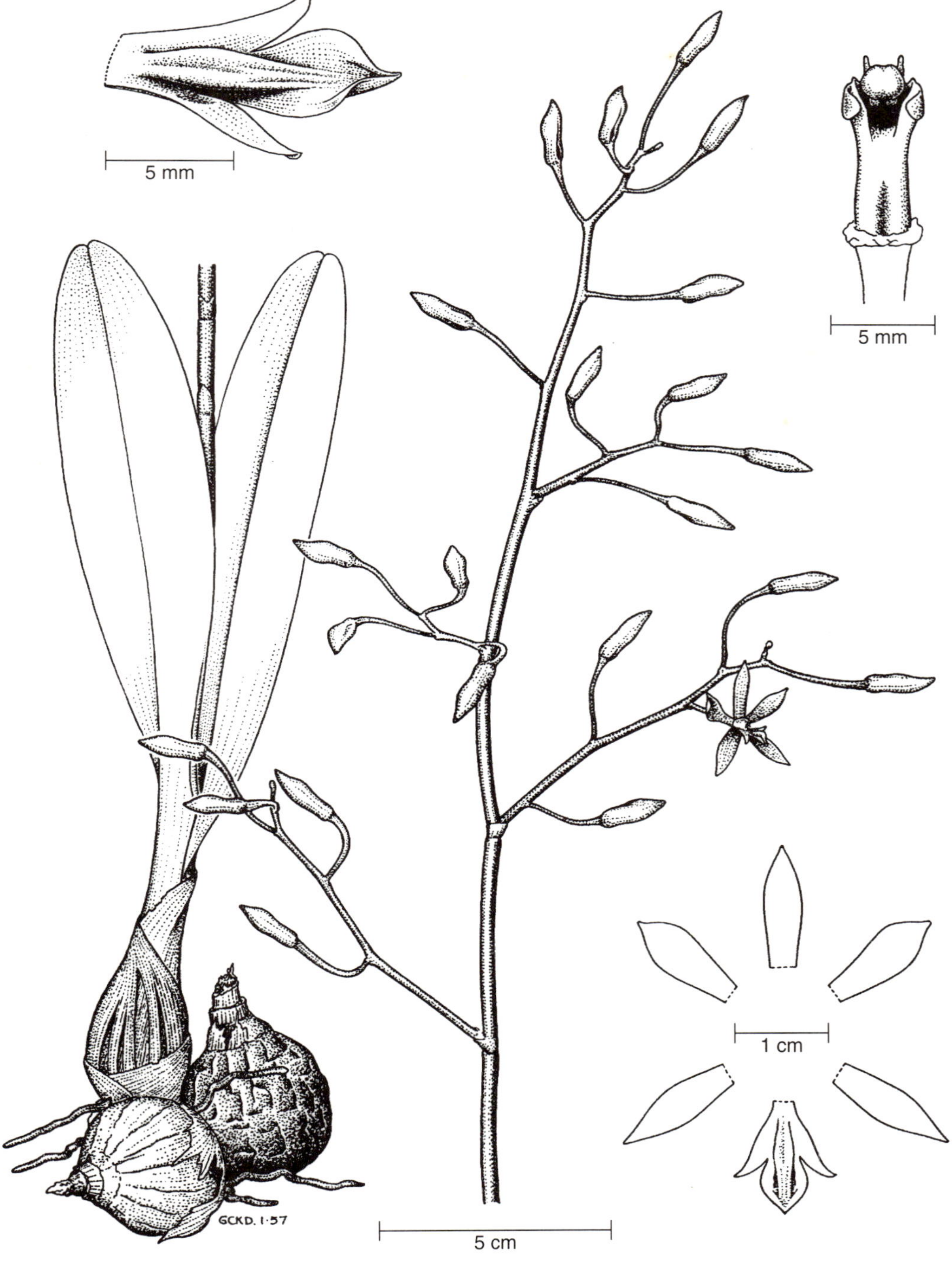

Figure 2-5. *Encyclia bradfordii.* Drawing #401 by G. C. K. Dunsterville. Please see text about alternative identifications.

The problem for us here is that the illustration of the flower in drawing #401 by Dunsterville better matches the lip pattern of *Encyclia isochila* (also found in Volume IV with a drawing), but the plant habit matches neither *Enc. isochila* nor *Enc. bradfordii.* Dunsterville's #401 has a tall panicle, a coarse flower stalk, wide obtusely tipped leaves, and wrinkled pseudobulbs. This many-flowered entity from Venezuela needs more research. We know so little about *Enc. bradfordii* and *Enc. isochila.* If they occur in continental South America, then my question about the identification may be unfounded. We could also be dealing with a taxon that is yet unnamed. Please see *Enc. sclerocladia* for additional information.

Bragançe Family Encyclia

Brazil

(FIGURES 1-1, 2-6; PLATE 9)

Encyclia bragancae Ruschi. 1976. *Bol. Mus. Biol. Mello Leitão* 84:1.

Synonyms

Encyclia xuxaensis Fowlie and Duveen. 1991. *Orchid Digest* 55(4):157.
Encyclia xuxiana Fowlie and Duveen. 1992. *Orchid Digest* 56(4):204.

The members of the Bragançe family were descendants of the Brazilian emperor Dom Pedro II, and they are commemorated by this namesake orchid. It is unique because of its frilly pink midlobe, strongly spatulate brown petals, and winter flowering season. Unfortunately for Fowlie and Duveen, it had been named before their discovery of a flowering plant at Orquidário Binot in 1991. Their name, *xuxiana,* in honor of Duveen's wife, will have to become a synonym. The lip is described as both crisped and fimbriate in its apical portions, apparent in the lack of detail in the drawing from Ruschi's publication of the species.

The flowers are fragrant, and the sepals are spatulate and yellow-green, about 8 × 20 mm. The petals are similar and are a little wider across the tips. The lip has lobes 15 mm wide, and the midlobe is 15 mm long and wide. The lateral lobes are yellowish with purple veining. The column is yellowish, 12 mm long, enlarged toward the apex, and has small auricles. The plant was found on the Pico de Bandeira in the Parque Nacional do Caparão at 2200 m elevation. Nothing more about it has been found. As is true of many of these encyclias, new populations should be raised from seed so that the numbers of plants in cultivation can be increased. This may be, after all, a major means of orchid conservation for many easy-to-grow species with interesting variations. No other *Encyclia* species has such a frilly lip, and I am certain that hybridizers will want to know about this particular characteristic.

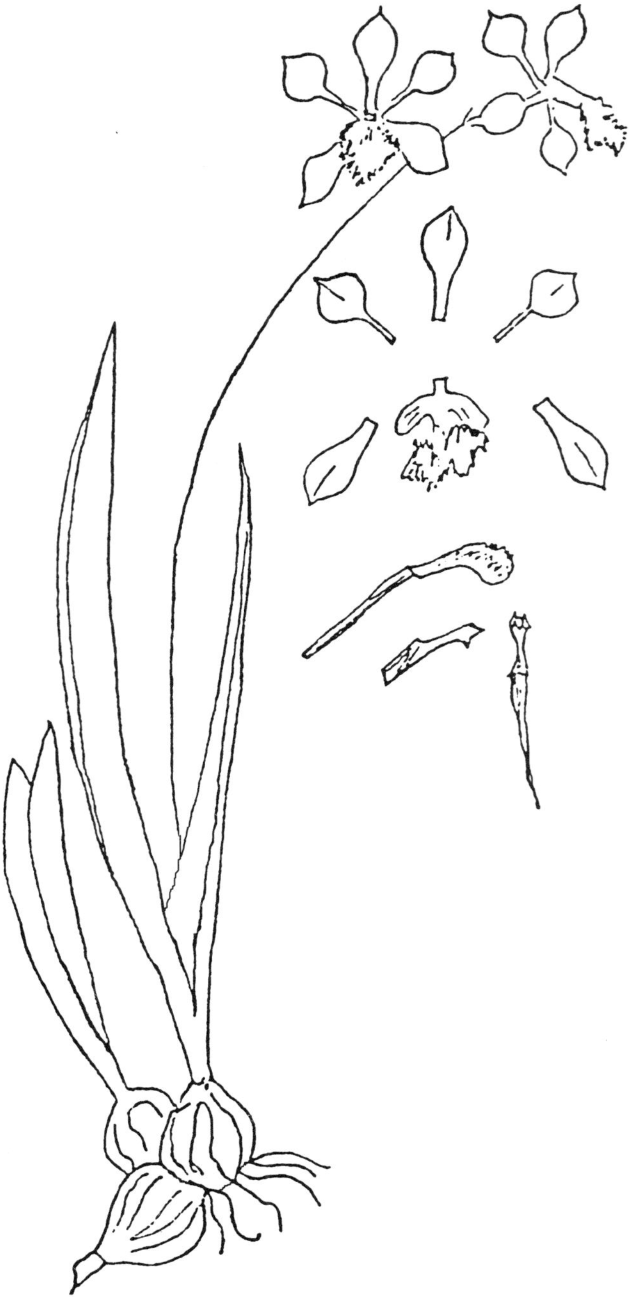

Figure 2-6. *Encyclia bragancae.* Type drawing of the species from the *Boletim Museu Biológico Mello Leitão.* The drawing depicts well the frilled and crisped nature of the lip midlobe that is unique in this genus.

Burle Marx's Encyclia

Brazil

(FIGURE 1-2; PLATE 10)

Encyclia burle-marxii Pabst. 1979. *Bradea* 2(48):313.

Another of the more recent Brazilian species to be described is this one dedicated to the famous imaginative and talented landscaper, garden and fountain designer, horticulturist, nurseryman, jewelry designer, and artist-sculptor, Roberto Burle Marx. Many of his outdoor works may be seen from Brasilia to Venezuela to Miami. He also grew and appreciated orchids, and it is appropriate that this Brazilian species honors his name. The plants were collected first by Burle Marx growing on rocks "along a seldom traveled road" (Fowlie and Duveen, 1992) near Baixo Guandu in northwestern Espírito Santo state not far from its mountainous border with Minas Gerais. They flowered later in his garden on 14 November 1978. The type specimen is in the Pabst Herbarium. Pabst states that the size of the flowers and the ovate midlobe of the lip distinguish this species from related forms in Brazil.

The type description tells us the pseudobulbs are ovoid, 5 cm tall × 2.8 cm wide, and three-leaved; leaves are linear-oblong, straight and rigid, with obtuse apices, and 20–22 cm long × 2.7–3.0 cm wide. The raceme is erect, has around 20 flowers in succession, and is about 30 cm long. The flowers are brown with a white lip that is veined in magenta. The dorsal sepal is lanceolate, 20–23 × 6–7 mm, and the petals are spatulate with obtuse apices, 22 × 7 mm. The lateral lobes are rectangular-ligulate in outline with rounded tips, 10 mm long × 5 mm wide. The ovate midlobe is 13 × 11 mm with a pointed apex, and the callus protrudes onto the disc with the central veins continuing to the apex. The column is straight and has auricles; the ovary is smooth with a sheen.

This species, like others that are poorly known, should be propagated from seed in Brazil for distribution around the world. Fowlie and Duveen (1992) found plants at Orquidário Binot from Espírito Santo state and compared them favorably with the type specimen. The plants flower in November and December, the Brazilian spring.

Cardim's Encyclia

Brazil

(FIGURE 1-2)

Encyclia cardimii Pabst and Mello. 1977. *Bradea* 2(27):183.

This species is dedicated to Flávio Affonso Cardim, the head orchid grower at Orquidário Binot for 40 years, friend and teacher about the plants. The authors say it might be rash to describe another rupiculous *Encyclia* species from the higher regions of central Brazil, but this one has different characteristics that make it easy to recognize. They point out the uniform tawny yellow-brown color of the sepals and petals, the cuneiform lateral lobes of the lip, and the callus rib that extends to the indented apex of the white lip with its magenta flushes and veining. The original plants were collected by Arturo Ferreira de Mello near Montes Claros in Minas Gerais state of Brazil. They were then grown in his collection at Nogueria near Petrópolis with the help of Cardim. According to Fowlie and Duveen (1992), Ferreira disappeared suddenly and his collection went to pieces for lack of care. It was not possible for them to trace any specimens of this species, and it presently is apparently lost to cultivation.

According to the type description, the pseudobulbs were narrowly ovoid, 8 cm tall × 2 cm wide. The three leaves were 16–19 cm long × 2.3 cm wide with obtuse apices. The raceme was about 15 cm tall and had four flowers. The sepals were broadly lanceolate, obtuse, measuring 18–20 × 5.5–6.5 mm. The petals were spatulate with broad tips, 16–18 × 7.0–7.5 mm. The lip had a cuneate base extending into the lateral lobes, 8 × 4 mm, with rounded tips reflexed in the expanded position. The midlobe was rounded, minutely apiculate, and 10 mm across. The column had auricles, was angled to the ovary, and 10 mm long. The pedicellate ovary was sparsely verruculose, 2.5 cm. The type specimen cited by Pabst is #EN-42A in his herbarium.

This species is like the other Brazilian species, so often with brown-green sepals and petals and white lips with magenta veining, growing in varied places and habitats, seldom seen, and often confused as a result of misidentifications. We have little more than the type descriptions to go on, and no photos nor illustrations. It is frustrating to try to separate these species in a meaningful fashion. By listing and describing them in this book, one could hope that new collections could ultimately be made and the forms compared. Since most of these species grow in rocky or mountainous habitats, their environments are often comparatively "safe" from destruction. We can hope that their endemic natures in obscure locations will preserve them for the future. The endemism is particularly pronounced in Brazil, which seems to have more of these *Encyclia* species than other South American countries. The geological and geographical histories of the various areas will no doubt provide further understanding and explanations of these distributions.

Caximbo Hill Encyclia

Brazil

(FIGURE 1-2; PLATE 11)

Encyclia caximboensis Menezes. 1992. *Orchid Digest* 56(1):10.

Another rock dweller, this species was first collected at 400 m in August, 1991, in the military reserve of Caximbo Hill in the Brazilian state of Pará. The collector was Antonio Prenholato, an orchid enthusiast from the Brasilia Orchid Society and an Air Force captain. The type specimen is conserved at the herbarium of the University of Brasilia.

The pseudobulbs are ovoid-oblongate, brownish green, 3–7 cm tall × 3–5 cm wide. The one to three leaves are erect, leathery, 13–22 cm long × 2.0–2.5 cm wide, with pointed tip. The inflorescence is 43–50 cm tall, a raceme with nine fragrant flowers. The sepals and petals are greenish with brown veining. The sepals are lanceolate, acute, 25 × 5 mm. The petals are subspatulate-oblanceolate, 17 × 5 mm. The labellum is white with a greenish base and the whole lip, including the lateral lobes, is covered with purple veining, even on the callus, except for a narrow white margin. The lateral lobes are broadly shouldered, obliquely triangular, and point down toward the midlobe. Their tips are obtusely pointed. The column is white suffused with reddish stains, slightly curved, and 12 mm long. Except for the type description, no other information is available on this species.

Horned-columned Encyclia

Colombia, Costa Rica, El Salvador, México, Nicaragua, Panama, Venezuela

(FIGURE 1-2; PLATE 12)

Encyclia ceratistes (Lindley) Schlechter. 1919. *Repert. Spec. Nov. Regni Veg.*, Beihefte 6:74.

Synonyms

Epidendrum ceratistes Lindley. 1844. *Bot. Reg.* 30, item 92.
Epidendrum ramonense Reichenbach f. 1866. *Beitr. Orch. Cent.-Am.*, p. 81.
Encyclia ramonensis (Reichenbach f.) Schlechter. 1918. *Beih. Bot. Centralbl.* 36(2):473.
Epidendrum oncidioides var. *ramonense* (Reichenbach f.) Ames, Hubbard, and Schweinfurth. 1935. *Bot. Mus. Leaflt., Harv. Univ.* 3:103.
Encyclia oncidioides var. *ramonensis* (Reichenbach f.) Hoehne. 1952. *Arq. Bot. Est. São Paulo*, n.s., 2:153.

As noted in Volume V of this series, *Encyclia ceratistes* is one of three or four species that is found in both Central America and South America. The geography must reflect on the age and evolution of these species. The species is named because of a pair of small hornlike projections from the sides of the column—not the quadrate larger auricles that are more typical in this genus. A diagnostic oval brown veining spot is located at the base of each lateral lobe, and the midlobe is purple-veined and creamy white, a pointed spade-shaped structure. The sepals and petals are starry and bright yellow-green. A complete discussion may be found in Volume V. The plant grows well in baskets or on cork slabs. Drawing #110 by Dunsterville (1979) shows this species well.

Chapada dos Viadeiros Encyclia

Brazil

(FIGURE 1-2; PLATES 13, 14)

Encyclia chapadensis Menezes. 1992. *Orchid Digest* 56(3):133.

Because of the construction of a new highway from Brasilia to Goias, the rock outcrops and giant boulders upon which this species has found its habitat have already been stripped by orchid collectors. Menezes considers the species on the edge of extinction. It is (was) found in the municipality of Alto Paraiso in Goias state.

The pseudobulbs are green-brown, oblongate, 6–7 cm tall × 2.0–2.5 cm wide. The two to three leaves are leathery, erect, 8–10 cm long × 2.0–2.5 cm wide. The flower scapes can be 100 cm tall simple unbranched racemes with 7–15 large fragrant flowers. The sepals and petals are red-brown. The sepals are lanceolate, pointed, 22 × 6 mm, and the petals are a little wider toward their tips, 21 × 7 mm. The lip is rose-flushed, the midlobe rounded, the lateral lobes almost rectangular. Purple veining is on the midlobe and the lower portions of the lateral lobes. The tips of the lateral lobes are not marked with veining, and the midlobe has a white margin.

Whenever a plant of this species is found in cultivation, it should be self-pollinated and seedlings raised for conservation of the species. Being a rock inhabitant, it should grow well on cork slabs in the bright greenhouse.

Green and White Encyclia

Brazil, Colombia, Ecuador, Guyana, Peru, Suriname, Venezuela

(FIGURE 1-2; PLATE 15)

Encyclia chloroleuca (Hooker) Neumann. 1845–1846. *Rev. Hort.*, ser. 2, 4:138.

Synonyms

Epidendrum chloroleucum Hooker. 1837. *Curtis's Bot. Mag.* 64, t. 3557.
Epidendrum chloranthum Lindley. 1838. *Bot. Reg.* 24:25, misc. 28.
Epidendrum chloroleucum var. *fuscoluteum* Regel. 1857. *Index Sem. Hort. Petróp.*, p. 29.
Encyclia thienii Dodson. 1989. *Icones Plantarum Tropicarum*, ser. II, fasc. 5, plate 458.

This fleshy, greenish and unmarked, small-flowered *Encyclia* with whitish lip and two or three leaves is distinguished by the heavy central rib extending from the callus to the apex of the lip. Also, it has definite auricles on the column, a large callus, and a smooth ovary. The leaves are described as "obscurely bilobed" at the apices. The species is widely distributed in South America. The original plant was sent by Richard Schomburgk to Loddiges and it flowered in March, 1838. Both Hooker and Lindley seem to have named it, but Hooker's epithet was published first and has the priority; nothing, however, has been recorded about that little competition. The type specimens for both species are in the Lindley Herbarium at Kew Gardens and came from Guyana. The plate in *Curtis's Botanical Magazine* gives a fine impression of the species.

My flower observations show the sepals 15 × 4 mm, lanceolate to slightly ovate, and pointed. The petals are obovate and slightly clawed, 15 × 4 mm. The lip is 15 mm long, and the lateral lobes are narrow and rounded, sometimes even pointed at the tip. The midlobe varies from spade-shaped and distinctly pointed to rounded with a slight tip; it is 5–6 mm wide × 6–7 mm long. The isthmus with the white callus is narrow and 2–3 mm long. The central vein often has a purple streak over it, sometimes also the lateral veins on either side. The lime green of the sepals and petals may also have a dark flush on some forms. The column is usually a little dark, and the anther cap is a bright yellow by contrast. Sometimes the sides of the midlobe reflex along the center rib. It would be interesting to know what pollinator is involved with this species. Most accounts say the flowers are not fragrant.

Eric Christenson has provided me with a photograph of the type specimen of *Epidendrum chloroleucum* var. *fuscoluteum,* three flowers on a short stalk, conserved at the Komarov Botanical Institute in St. Petersburg, Russia. *Encyclia acuta* is another likely synonym of this species, and, according to Castro (1998), *Enc. viridiflava* is also. Cogniaux describes it only with two words—flower yellow-brown—and then mentions México as its possible source. Without more information it is impossible to tell what the Mexican plant really might be, but *Encyclia chloroleuca* does not grow in Mesoamerica.

Shell-lipped Encyclia

Brazil, Guyana, Paraguay, Venezuela

(FIGURES 1-2, 2-7; PLATE 16)

Encyclia conchaechila (B. Rodrigues) Porto and Brade. 1935. *Rodriguésia* 1:2.

Synonyms

Epidendrum conchaechilum B. Rodrigues. 1877. *Gen. & Spec. Orch. Nov.* 1:53.
Epidendrum linearfolioides Kraenzlin. 1911. *Kungl. Sv. Vet. Akad. Handlingar* 46(10):55.
Encyclia bicornuta Brade. 1943. *Arq. Ser. Forest., Rio de Jan.* 2(1):5, pl. 4.
Encyclia amicta subsp. *conchaechila* (B. Rodrigues) Carnevali and Ramírez. 1988. *XLVI Exposicíon Orquídeas, Comité de Orquideología,* Bol. No. 23:36.

Misapplications (*auctor non*)

Epidendrum bicornutum Hooker. 1856. *Bonplandia* 4:327 = *Caularthron bicornutum* (Hooker) Bentham, 1881, *Journ. Linn. Soc.* 18:312. Also see *Curtis's Bot. Mag.,* t. 3332, 1834.
Epidendrum bicornutum Hooker = Legume species. Note in *Floral Cabinet* 2:182, 1844.

I too have waffled on the position of this species, but I have finally decided to separate it from *Encyclia amicta* (please see above) and the others of its complex. It grew in northern Brazil in Alto Amazonas on the rivers Solimoes and Negro, so it is not surprising to find it reported also in Venezuela. The plants are comparatively small with pyriform pseudobulbs and two narrow, pointed leaves that may reach 30 cm long × 8 mm wide. The flower stalks are shorter than the leaves and may be either simple or branched.

The taxon was imaginatively named because the concave midlobe of the lip appeared like a scallop shell with its rounded shape and the five main raised veins looked like the ribs of the shell. The central main vein runs to the pointed apex of the midlobe, and the two major veins on either side of it radiate laterally on the disc. All five show verrucosities, especially toward their lightly keeled bases where they develop from the forcipate portions of the callus. The ovary surface is roughened or asperous.

The lanceolate, pointed sepals and petals are yellow-brown, and the lip is yellow. Measurements include sepals at 15 × 3.5 mm, with the petals

Figure 2-7. *Encyclia conchaechila.* Type drawing by Barbosa Rodrigues published as *Epidendrum conchaechilum* by Cogniaux. Plate 13 of *Martius* 3(5).

a little shorter. The lip is 15–17 mm long, and the lateral lobes are pointed. The column is straight, slim at the middle and widened toward the base and apex. It has no auricles, but the edges are slightly winged. The anther cap is deeply grooved at the top with the two "horns" darkly colored in their apical portions. Although not mentioned in the type description of Barbosa Rodrigues, nor by Cogniaux in *Martius*, the species is distinctive in having small hairs covering the basal portions of the callus, and they are matched by the roughness of the ovary. Cogniaux noted that he had not seen the actual specimen but had obtained his information directly from Rodrigues. These hairy flower parts, also observable in a few other species with darker anther cap color, may indicate a different pollinator for these species. Nothing has been said about possible fragrances or pollinators.

Confusing Encyclia

Brazil

(FIGURE 1-2; PLATE 17)

Encyclia confusa Menezes. 1991. *Orchid Digest* 55:132.

Menezes, in publishing this species, describes how by chance she was sent two specimens by different people who were asking for identifications. Nazir Pinheiro Perex had a plant from Espírito Santo state labeled *Encyclia odoratissima*. Delco Jorge dos Santos had a plant, also from Espírito Santo, that had been collected from a fallen tree near the Pedra Azul region at about 800 m. Upon examination, Menezes decided that the Pinheiro plant differed from *Enc. patens* (synonym *Enc. odoratissima*) on five different points. First, these flowers are larger. Second, the inflorescence is a simple raceme, not a panicle. Third, the petals are spatulate-oblanceolate with pointed apices, while those of *Enc. patens* are distinctly spatulate with obtuse apices. Fourth, the lateral lobes are long and winglike, while those of *Enc. patens* are shorter and more right-angled to the base of the lip. Fifth, the midlobe of the lip is oval, white, and has purple veining, while the midlobe of *Enc. patens* is yellow, rounded, barely striped at all, and reflexed on the sides of a prominent central rib.

The white lip and the oval shape of the midlobe, longer than broad, make this species quickly recognizable. Another species with a similar oval midlobe—not transverse but upright in orientation—is *Encyclia argentinensis;* this characteristic is not all that common among encyclias.

The pseudobulbs are green, ovoid, and 5.5–6.0 cm tall × 2.0–2.5 cm wide with one to two erect leaves 27–29 cm long × 2.0–2.2 cm wide. The scape is to 50 cm and has nine flowers with a slight fragrance in the type specimen. The sepals and petals are green with brown veining and flushes, lanceolate, and 24 × 6 mm. The petals are spatulate-oblanceolate, 22 × 5 mm. The midlobe has slightly undulate edges, and the column is yellow-green, slightly curved, 12 mm long. No mention is made of auricles or the surface of the ovary. The type specimen is conserved at the herbarium of the University of Brasilia.

Large-lipped Encyclia

Brazil?, Venezuela, Colombia, Central America, French Guiana, Guyana

(FIGURE 1-2; PLATE 18)

Encyclia cordigera (Kunth in Humboldt, Bonpland, and Kunth) Dressler. 1964. *Taxon* 13:247.

This relatively common species in northern South America, particularly reported in Venezuela, is questionably found in Brazil, perhaps in western areas north of the Amazon River. Please see the full account of the nomenclature and history of *Encyclia cordigera* in Volume V. The flowers are among the largest in the genus, and the lips can measure from 30 to nearly 50 mm long. Measurements vary, but a large, typical form can be as follows: dorsal sepal 34 × 11 mm, petal 33 × 9 mm, lip 45 mm, lateral lobe 18 × 5 mm, and midlobe 35 mm long × 40 mm wide. To my knowledge, no specimens have been found south of the Amazon River.

Sword-leafed Encyclia

Bolivia, Brazil, Ecuador, Peru

(FIGURE 1-2; PLATE 19)

Encyclia cyperifolia (Schweinfurth) Carnevali and Ramírez. 1993. *Monogr. Syst. Bot.* 45:1257.

Synonyms

Bletia ensiformis Ruiz and Pavón. 1798. *Syst. Veg. Fl. Peru et Chile* 1:230.
Epidendrum microtos var. *grandiflorum* Schweinfurth. 1943. *Bot. Mus. Leaflt., Harv. Univ.* 11:98.
Encyclia microtos var. *grandiflora* (Schweinfurth) Hoehne. 1952. *Arq. Bot. Est. São Paulo* 2(6):152.
Epidendrum cyperifolium Schweinfurth. 1953. *Bot. Mus. Leaflt., Harv. Univ.* 16:10.

Misapplication (*auctor non*)

Encyclia ensiformis (Vellozo) Hoehne. 1952. *Arq. Bot. Est. São Paulo* 2(6):2151 = *Epidendrum ensiforme* Vellozo. 1827. *Fl. Flum.* 9, t. 2 for plate. Text 1881, L. Netto, ed. *Arch. Mus. Nac. Rio de Janeiro* 5:356.

The original *ensiformis* epithet for this species refers to the leaves—like a long, narrow, two-edged, and pointed sword. Schweinfurth implied the same thing when he called it *cyperifolium*, meaning sedge-leafed. The leaves, compared to most others, are narrow for this genus, and the pseudobulbs are unusually small, squat, and round for the size of the plants. When I first saw a plant growing in Fred Fuchs's Nursery in Naranja, Florida, in April, 1985, I easily recognized it as something that orchid hobbyists had never seen before. Fred said, "Oh, we collected it near the airport while waiting for our flight [Bolivia to Lima]. Tell me what it is." After my return to New York, I found the plant matched the Ames photo of Ruiz and Pavón's type specimen in the Madrid herbarium. Charles Schweinfurth had penned the following on the sheet with the photo, "Lack of floral details makes identification impossible. It may be referable to *E. thrombodes* Reichenbach f.—(later amended)—to *E. cyperi-*

folium Schweinfurth." Schweinfurth had realized that there were homonymic problems with the original name, and that the *microtos* var. *grandiflorum* epithet also had problems.

The Peruvian type specimen of *Bletia ensiformis,* Ruiz and Pavón #29525, is in the Madrid herbarium. Having worked with some of the Brazilian encyclias, I too realized the homonymic problem with *Epidendrum ensiforme* Vellozo: the Ruiz and Pavón *Bletia ensiformis* was transferred to the genus *Encyclia* by Hoehne after Vellozo had used the same species epithet for *Epi. ensiforme,* a completely different Brazilian species. Carnevali and Ramírez transferred the Ruiz and Pavón *ensiformis* to *Encyclia,* and now we are all straight. The Vellozo *ensiforme* is a synonym of *Enc. oncidioides* (see below).

The flowers of *Encyclia cyperifolia* are an unusual combination of a rosy lilac color and yellow and are well described by Bennett and Christenson (1995), "Sepals and petals pale yellow tinted rose on the dorsal surface, the labellum midlobe yellow around the border, center tinted rose, the veins darker, the lateral lobes pale yellow, the column very pale green, the anther (cap) bright yellow." The dorsal sepal on my specimens runs to 18 × 3.5 mm, the petals 17 × 4 mm, the lip 17 mm long and 10 mm across the flattened pointed lateral lobes and 8 mm across the midlobe. The sepals are oblong-oblanceolate, the petals more spatulate. Lips are clawed at the base for about 3 mm that extends into the wedgelike triangular shape of the short lateral lobes. The isthmus with its broad, fleshy, creamy-colored callus is also extended into a long wedge-shaped base for the oval to circular midlobe. The midlobe is slightly indented at the tip and somewhat reflexed along the sides—the whole lip being of a distinctive shape. The ovary surface is rather papillose (to smooth), and the rough and coarse flower stalk is branched and extends only to the level of the leaf length. Pale yellowish edges coming to a broad point, more than a distinct auricle, distinguish the unusually upwardly curved column that is 10 mm long.

The plants grow in warm to hot mountain forests. They flower in the summer. Overall, the plant and flowers are so distinctive that one wonders why there have been so many problems concerning the plant's identity. Communication seems to be the answer.

Two-colored Encyclia

Brazil, Suriname

(FIGURES 1-2, 2-8; PLATES 20, 21)

Encyclia dichroma (Lindley) Schlechter. 1914. *Die Orchideen,* ed. 1, p. 209. Also an isonym transferred by Porto and Brade, 1935, in *Rodriguésia* 1(2):29.

Synonyms

Epidendrum dichromum Lindley. 1843. *Bot. Reg.,* 29: misc. 119, p. 78.
Epidendrum brasiliense Hawkes. 1956. *Orquídea* 18:174.
Encyclia dichroma subsp. *biflora* (B. Rodrigues) Fowlie. 1992. *Orchid Digest* 56(3):141.

Misapplication

Epidendrum biflorum B. Rodrigues. 1877. *Gen. & Spec. Orch. Nov.* 1:52. *Auctor non* Forst f., 1786 = *Dendrobium biflorum* (Forst f.) Swartz, Ruiz and Pavón, 1798 = *Epidendrum peruvianum* Hawkes, or Cogniaux, 1902 = *Epidendrum paranaense* B. Rodrigues (see *Bull. Herb. Boiss.*, ser. 2, 2:337, fig.).

This species is famous for its color and shape and for growing under dry conditions on rocky outcrops in Bahia state where it can form large masses coating the substrate. Fowlie (1992) has written and pictured its habitat, especially of the variety he calls *biflora*. Veitch says that the plants grow along rivers, establishing themselves on low, straggly bushes and sending their roots down into the sand. It is one of the two encyclias with rosy pink sepals and petals and darker lips, and was the only one in Brazil of that color until a very closely related species, *Encyclia ghillanyi* (now *Enc. quesneliana*), was named a few years ago.

Many still do not believe that they are two species, but think that they represent more than the usual variability in a single species. The number of synonyms for the complex, however, gives a clue for unraveling the great variability. For those who are convinced that the two species are separate, please consult Figure 2-31. In a few words, *quesneliana* is distinguished by larger flowers, wider, almost round petals, squared-off tips of the lateral lobes, lateral lobes pointing down when the lip is flattened to about 45 degrees, and more paniculate inflorescences. Sometimes it seems a match or the distinction is easy; other times the flower in question can match on some points and not others. The drawings (Figures 1-2, 2-8) are based on clones of just *dichroma* flowers. Such types show a reasonably continuous range in color. The *ghillanyi* variations, apparent in Figure 1-31, seem to be different enough—large truncate lateral lobes and almost circular petals—to distinguish it as a separate species. One complication, however, is that the earliest varietal name for the *dichroma* variant then provided an earlier basionym for a proper species name, *Encyclia quesneliana* (please see below), instead of *Enc. ghillanyi* for the congener.

The *Encyclia dichroma* plants are described as strong types with pseudobulbs to 10 cm long, of globose to conelike shape, and bearing as many as three tongue-shaped blunt leaves that may reach 25–30 cm long. Most of the plants I have seen in cultivation, however, tend to be smaller. Some smaller plants produce only one or two flowers on a stalk and have been called *Enc. biflora*, but others are more vigorous and may produce branched panicles considerably larger with many flowers. The flowers can vary from white through pale pink to a rosy red, the lip being darker to match the increased intensity of sepals and petals. Lips tend to be white edged and have a yellow area at their base. The plants flower in early summer in cultivation, or from September to October in the Brazilian spring. Flower details show sepals and petals may vary from elliptic to spatulate-rounded in shape and may appear almost like a miniature labiate cattleya in outline. Some clones may have petals that are nearly round and overlapping the dorsal sepal, and these are now recorded as plants of *Enc. quesneliana*.

Cogniaux in *Martius* gives sepal measurements as 22–25 × 7–9 mm. The petals are 21–23 × 11–13 mm. Lips are 21–24 mm long with broad rounded lateral lobes and a somewhat kidney-shaped midlobe with a

Figure 2-8. *Encyclia dichroma.* Type drawing by Barbosa Rodrigues published as *Epidendrum biflorum* and then transferred to *Epi. dichromum* var. *biflorum* by Cogniaux. Adapted from Plate 19 of *Martius* 3(5).

small notch at the apex. The callus extends onto the disc of the lip from the isthmus, but the veins are not particularly raised, though they may show as a darker striated pattern. The column is about 1 cm long and has distinct auricles. The ovary surface is smooth as are the flower stalks.

Plants grow best under drier, brighter conditions than those in which cattleyas grow, so do well in small hanging baskets or on slabs of cork. They should be kept in a bright airy location with some humidity still remaining in the evening. I have found them difficult to reestablish after importation, and they may just sit for a season with no new growths appearing. One must have patience. I have found this species particularly attractive and have often wondered why more hybrids with it have not been tried. The color and shape of good forms of the species (or perhaps they are plants of *Encyclia quesneliana* with nearly orbicular petals) are ideal, and from what I have seen, these characteristics show up well in the offspring. One of the best *Encyclia dichroma* hybrids is Royal Purple made by Bartley Schwarz of Highland Tropicals in California. The other parent was *Enc. atrorubens* from México.

Daytime Encyclia

Colombia, Suriname, Venezuela

(FIGURE 1-2; PLATE 22)

Encyclia diurna (Jacquin) Schlechter. 1919. *Repert. Spec. Nov. Regni Veg.*, Beihefte 6:74.

Synonyms

Limodorum diurnum Jacquin. 1790 (1791). *Collect. ad Bot., Chem., et Hist. Nat.* 4:107. Also cited is Jacquin. 1793. *Icones Pl. Rarum* 3, t. 603.

Cymbidium diurnum (Jacquin) Swartz. 1799. *Schrad. Journ. Bot.* 1:221.

Epidendrum diurnum (Jacquin) J. L. M. Poiret. 1810. *Encyc. Méth. Bot.*, Suppl. I, p. 375; also *Angrec diurne* J. L. M. Poiret.

Cymbidium glandulosum Kunth in Humboldt, Bonpland, and Kunth. 1816. *Nov. Gen. & Spec. Pl.* 1:340.

Epidendrum wageneri Klotzsch. 1851. *Otto and Dietr. Allgen. Gartenzeit.* 19: 250.

Epidendrum diurnum (Jacquin) Reichenbach f. 1866. *Beitr. Orch. Cent.-Am.*, p. 81.

Encyclia wageneri (Klotzsch) Schlechter. 1919. *Repert. Spec. Nov. Regni Veg.*, Beihefte 6:75.

Encyclia glandulosa (Kunth in Humboldt, Bonpland, and Kunth) Ortíz. 1991. *Orquideología* 18(1):99.

Misapplication (*auctor non*)

Encyclia virens Schlechter. 1914. *Die Orchideen*, ed. 1, p. 212.

The type specimen for this species was apparently collected in the valley of Caracas or on nearby Mount Avila north of the city, flowering in May and June. It is one of the first *Encyclia* species to have been described, having been a *Limodorum*, a *Cymbidium*, and an *Epidendrum* before Hooker described *Encyclia* in 1828 with *Enc. viridiflora* from Brazil. Having had its share of debate and confusion with various other species, plants of *diurnum* or *diurna* have been misidentified as *Enc. oncidioides, Enc. selligera, Enc. ochrantha, Enc. plicata, Enc. gracilis, Enc. caicensis, Enc. altissima, Enc. acuti-*

folia, and *Enc. virens.* These misidentifications have falsely increased the taxon's range considerably, and now we can straighten out the problem. It seems to be endemic to Venezuela and possibly Colombia and Suriname.

The species has also been called the Tall Encyclia because of its ample flower panicles that can reach over a meter. Dunsterville describes the plants with panicles 2 m long and branches 50 cm, the stalk brown or maroon. The pseudobulbs may reach 12 cm long and the leaves 50 cm long × 3 cm wide, putting the species in a size class with *Encyclia oncidioides* and *Enc. pachyantha.* The plants are usually two-leaved, but three leaves are not unusual, and the pseudobulbs cluster together. The species is epiphytic and grows at elevations as high as 1050 m.

The flowers are light lime green on the sepals and petals, sometimes having a purplish flush or veining, especially on the sepals. The sepals measure 23 × 7 mm, are obovate and pointed. The petals are wider, 23 × 10 mm, also obovate but with bluntly pointed apices. The lip measures 20 mm long and 22 mm across the lateral lobes. The midlobe is 12 mm across, rounded, and white with purple veining on its disc. The base of the lip is greenish and veined about halfway toward the more narrowed rounded white tips. The five major veins of the midlobe are slightly thickened. The 10 mm column is greenish with purple flushes and has a yellow anther cap and small auricles. The ovary surface is roughened. The plants flower in summer.

A few hybrids with *Epidendrum diurnum* as one parent have been registered with the Royal Horticultural Society. Because of many misidentifications, particularly with Bahamian and Cuban encyclias, one cannot be certain that the registered parentages are accurate. They should, when possible, be checked if the *diurna* parent plant still exists. Drawing #109 by Dunsterville (1979) shows this species well.

Doering's Encyclia

Brazil

(FIGURE 1-3; PLATE 23)

Encyclia doeringii Hoehne. 1946. *Arq. Bot. Est. São Paulo* 2(2):24, t. 11.

Synonym

Epidendrum doeringii (Hoehne) Hawkes. 1956. *Orquídea* 18:175.

Encyclia doeringii comes from eastern Paraná in southwestern Brazil, a region so distant from areas north of the Amazon that I have always wondered how *Enc. doeringii* could really be a synonym of *Enc. cordigera.* Hoehne's drawings of *doeringii* certainly look similar to *cordigera*, but that will not account for the distribution gap, and the two species have been confused.

The drawings show the pseudobulbs round and squat with three broad, obtuse leaves held somewhat horizontally. The flower scape is shorter than the leaves. The sepals are oblanceolate-ovate, and the petals spatulate with a tapering claw and undulate edges. The lateral sepals are shown with inturned tips like those of *Encyclia cordigera*, but not the tip of the dorsal sepal or the petals. The lateral lobes of the lip are rounded and flared upwardly. The midlobe is more orbicular than elongated like that of *Enc. cordigera,* emerges from a broad isthmus, and is ruffled a little around the edges coming to a point. For a Brazilian *Encyclia* species these are all small details that make me doubt if we have *Enc. cordigera* at hand. The column is without wings or auricles, is widest across the stigmatic area, and slightly S-curved in side view.

A photocopy of the type specimen shows the dorsal sepal to be obovate-lanceolate, acute, 30 × 10 mm. The lateral sepals are oblique obovate-lanceolate, acute, 33 × 9 mm. The petals are spatulate with a wedge-shaped claw to a narrow base, obtuse at the tips, 30 mm long × 14 mm across the widest portion, and undulate edged. The lip is 30 mm long with a 2 mm clawed-cuneate base producing 15 mm lateral lobes with parallel sides and rounded apices sweeping down and then up at the tips, 4–5 mm wide. The isthmus is 5.5 mm wide and about 3 mm long to form the midlobe, which is 18 mm long × 19 mm wide. The column is 14 mm, and the pedecillate ovary 27 mm long.

We should have another look at this species before we jump to further conclusions. The type sample was sent to Hoehne by Ricardo Doering of São Paulo in 1942, and there is only a card with a dissected flower, #52150, at the Instituto de Botanica. Hoehne's signature is on the label with the date of the description, October 28, 1944. We have to consider *Encyclia doeringii* a separate species, and a new collection or two with habitat data could resolve the lingering problems. I have never heard of this species in cultivation, nor talked to anyone who had seen it.

Dutra's Encyclia

Brazil

(FIGURE 1-3; PLATE 24)

Encyclia dutrai Pabst. 1955. *Arq. Bot. Est. São Paulo* 3:110, t. 28.

Synonym

Epidendrum dutrai (Pabst) Hawkes. 1956. *Orquídea* 18:175.

Coming from Torres, near Morro do Forno, in Rio Grande do Sul state in Brazil, this species has only been collected once—when the type description was written by Pabst. The specimen, #1164, had been called *Encyclia gracilis* in an unpublished manuscript by João Dutra, whose drawings were used to illustrate the Pabst paper on the species. The notes tell us that the sepals and petals were greenish suffused with purple, and the lip was white with red-brown veins. The drawing shows a round pseudobulb and two stiff semi-upright leaves. The sepals and petals are spatulate-obovate with narrower bases and irregular wavy margins and reflexed tips or curls on some of the perianth segments. The type plant also shows an inflorescence shorter than the leaves with three flowers. The lip is shouldered and has downswept lateral lobes with rounded tips and a midlobe with a cuneate base and transversely oval midlobe. The lip margin is crenate or undulate, and the midlobe reflexes back on either side of the center vein. The column has auricles, is wasp-waisted about the middle, and is broadest at the apex. Castro (1998) says *Enc. dutrai* flowers are the same as those of *Enc. pauciflora,* but this hardly seems possible considering their measurements and locations in Brazil.

If anyone has had a more recent view of this species, a picture or two should be taken to make a report of the finding. As is true for many other Brazilian encyclias, we need to know more about distribution, habitat, and conservation status. Are these species so rare and endemic that they have only been seen once in the wild and then never again? The number of herbarium sheets around the world on a given species may or may not reflect its occurrence in a given location. Often, however, with only a single specimen to consult for study, there are not enough open flowers to see important details—or they are so tightly glued to the paper they may be useless. Writing a set of descriptions such as these in an attempt to monograph the genus leads to many frustrations. I can only wonder how many of these species will turn out to be valid taxa in the future and not just poor records on other poorly known species. I am mostly wondering, however, how to explain what appears to be a great number of species known from only small habitat areas. Is it the geography or the ecology of the habitat area? Or will we ever know? Critical habitat information and adequate specimens will be the answer for herbarium conservation after the actual habitat is destroyed.

Duveen's Encyclia

Brazil

(FIGURE 1-3; PLATE 25)

Encyclia duveenii Pabst. 1976. *Bradea* 2(12):66.

Denis Duveen was a long-time Brazilian orchid enthusiast, a respected friend of many orchid growers, a collector of orchids and orchid books, and a writer about them. He was also a field-worker with many discoveries and observations to his credit, and an expert on perfumes, essential oils, and their fragrances. It is appropriate that Guido Pabst, an outstanding orchid taxonomist of Brazil and a director of Varig Airlines, has chosen to commemorate him with this *Encyclia duveenii.* Pabst writes, "This new species is readily recognized in its dark brown sepals and petals which are bent backward, often to a parallel position with the ovary. The lip is of intense magenta color." The type plants were found near Grão Mogol in Minas Gerais state, flowering in cultivation in 1973.

The pseudobulbs are narrowly conical, 5–6 cm tall, with two leaves, linear oblong with rounded apex, 15 cm long × 1.5–1.8 cm wide. The scape is 65 cm tall, an open panicle with about 20 flowers, the individual branches with four to five flowers. The dorsal sepal is "obscurely subspatulate," 18 × 4.5 mm, and the petals are spatulate, 18 × 5.5 mm. The lip has a rounded midlobe with slightly raised, non-verrucose veins, and with lateral lobes that are oblong and point downward. The column is 10 mm long and biauriculate, and the ovary is 1.5–2.0 cm long with the pedicel and angled to the column.

Plants grow well in a hanging basket with osmunda, and they would probably also do well on cork slabs. The flowers will be known for their rich coloration, but, unfortunately, not for their configuration of reflexed sepals and petals.

Edith's Encyclia

Brazil

(FIGURE 1-3; PLATE 26)

Encyclia edithiana Menezes. 1996. *Bol. C.A.O.B.* 26:22.

This species is dedicated by Lou Menezes to Edith Marinelli Oliveira, orchid enthusiast of São Paulo, Brazil. It was found growing epiphytically along the river Madeira in Amazonas state of Brazil, flowering in September and October, the Brazilian spring. It had a distinctly ornamental appearance with its numerous fragrant flowers in panicles 70 cm long. The sepals and petals were yellow-green veined and spotted toward the apices with red-brown, and the lips were white almost completely covered with bright rose-pink veining as can be seen in Plate 26. The white edge of the midlobe sets off the beautiful contrast between the veining and the lip color. The holotype is conserved at the herbarium of the University of Brasilia.

The type measurements are as follows. Pseudobulbs are 5–7 cm tall × 3.3–3.6 cm wide, with two lanceolate leaves, 22–23 cm long × 2.0–2.4 cm wide. The sepals are ovate-lanceolate, and the petals are ovate-spatulate, 16 × 5 mm. The lip has ligulate lateral lobes, and the midlobe is oval to rounded. The column is biauriculate, 7 mm long.

Expanded Encyclia

Colombia

(FIGURES 1-3, 2-9)

Encyclia expansa (Reichenbach f.) Ortíz. 1995. *Orquideología* 18(1):99.

Synonym

Epidendrum expansum Reichenbach f. 1877. *Linnaea* 41:115.

Encyclia expansa is very close to *Enc. ceratistes,* according to what Garay has written (1973). It was re-collected by Gilberto Escobar in Colombia, Department of Antioquia, for the first time since Reichenbach's 1877 description of the species. The new specimen is conserved at the Ames Herbarium as Escobar #673 without a more exact locality and was cultivated in Medellín. A color photo in *Orquideología* (1973, 8[2]:100) is our only record on this species besides the type description. Like *Enc. ceratistes, Enc. expansa* has a medium-sized flower with greenish oblanceolate-obovate sepals and petals. It has a white lip with red-purple veining, a spadelike midlobe shape, and a pointed tip with a rufflelike curl on either side. The flower also appears in the color photograph to have auricles on either side of the column apex, though the Reichenbach type sheet #994 with a sketch of the flower details shows a bare column without auricles. Also, the column is not mentioned in the type description. If the column

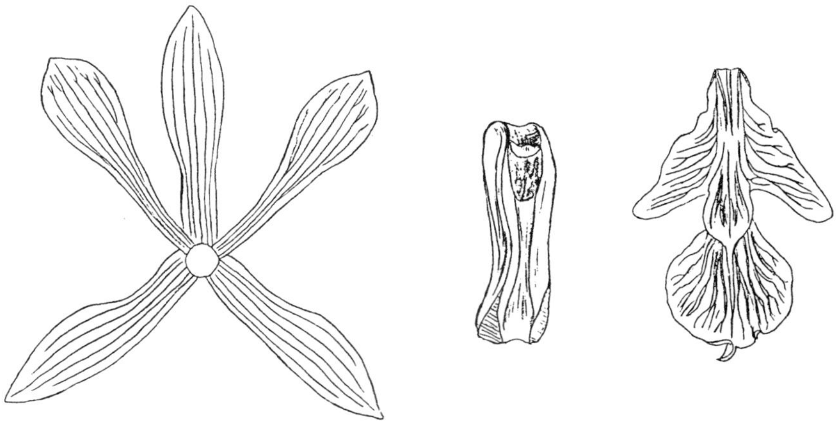

Figure 2-9. *Encyclia expansa.* A flower from the type specimen #994 of *Epidendrum expansum* collected in Colombia and conserved in the Reichenbach Herbarium in Vienna. Camera lucida drawing by Leslie Garay.

had auricles, the plant's identification would be in some doubt as the absence or presence of auricles is a characteristic flower detail that does not change much nor vary within a given species population.

The column of *Encyclia ceratistes* has two narrow hornlike auricles that, in fact, account for its name as the Horned-columned Encyclia. Also, the lip lateral lobes of *Enc. ceratistes* have a characteristic hallmark oval brown patch at their bases that this flower does not have. The lip of *Enc. expansa* has much more of the colored veining than does the lip of *Enc. ceratistes*. The differences are fairly distinctive between these two species, and therefore I am inclined to say they are completely different and not too closely related.

All this aside, we need more of *Encyclia expansa* to study and to ascertain its true status. Meanwhile it must be considered in its own right as a separate colorful species that is poorly known and poorly represented in the herbaria of the world.

Yellow-lipped Encyclia

Brazil, Venezuela

(FIGURES 1-3, 2-10; PLATE 27)

Encyclia flava (Lindley) Porto and Brade. 1935. *Rodriguésia* 1(2):29.

Synonyms

Epidendrum flavum Lindley. 1841. *Hooker Journ. Bot.* 3:83.

Epidendrum tripartitum Vellozo. 1881. *Fl. Flum.* (text for vol. 9); for plate alone, 1827 (1835), in *Fl. Flum.* 9, t. 7.

Epidendrum mapuerae Huebner. 1909. *Bol. Mus. Goeldi (Museu Paráense)* 5: 329.

Encyclia tripartita (Vellozo) Hoehne. 1943. *Arq. Bot. Est. São Paulo*, n.s., 2(6):151.

shape and orientation. Those two do not fit together, nor does *Enc. flava* fit with *Enc. conchaechilum*. I feel it is better for the time being to keep this species as a separate taxon. Other reports of its occurrence have placed it in Rio de Janeiro state, Minas Gerais, Goias, and subsequently in Venezuela by Dunsterville's drawing #258.

This *Encyclia* species is also again reported from Venezuela through the work of Carnevali et al. (1994) on another specimen, drawing #180 by Dunsterville. The authors write that there has always been confusion about the proper name for this species. Although they cite *Encyclia amicta* (Linden and Reichenbach f.) Schlechter as an alternative, they reject it because of its "obscurity" in Venezuela. It, also, is not a match. The species does seem, however, to match Vellozo's *tripartita*. It may also match *Enc. mapuerae* from Amazonas state, and I have placed that in synonymy for the time being. Obviously, it is a problem species and needs more research. The flowers are described as being pale yellow, about 3.7 cm in diameter. The lateral lobes are prominent, triangular, and pointed. The ovary surface is smooth. The leaves are about 30 cm long, and the somewhat short inflorescence is usually panicled only at the base.

Fowlie's Encyclia

Brazil

(FIGURE 1-3; PLATES 28, 29)

Encyclia fowliei Duveen. 1990. *Orchid Digest* 54(1):38.

Synonym

Encyclia bahiensis Menezes. 1989. *Brasil Forestal Pesquisa* 68:55–56. *Nomen nudum.*

The amazing things about this species are the flowers with their fascinating tessellations in red-brown veining on a yellow-green background, the very large petals, and why it was never discovered earlier in the many botanical explorations of Brazil. No other species in the genus has remotely similar flowers.

The plants were found in Bahia state growing southeast of Itapetinga. They were flowering in October and November, the Brazilian spring. Lacking a type specimen in flower to dry, Menezes substituted a color photo with her type description as a lectotype for the species, but that, unfortunately, would not stand. The type specimen for the *fowliei* epithet was deposited by Duveen in the UCLA Herbarium in Pasadena, California, and that does stand. Aranda Orchids in Teresópolis provided the specimen from collections they had made in Bahia.

Duveen's article provides us with most of the botanical details. These epiphytes grow to about 30 cm tall with reddish pseudobulbs and heavy roots. The rounded pseudobulbs are 2–3 cm tall and bear one or two leaves to 32 cm long × 2 cm wide. The flower scapes are to 30 cm tall, red-brown in color, and unbranched, and bear five to six flowers. The heavy flowers nod and are 4.5–5.0 cm in spread. The sepals measure 22 × 12 mm and tend to be curled in natural position. When flattened, they are obovate-spatulate with narrowed bases and obtusely pointed tips. The petals are 20 × 20 mm, the lower third to half being a comparatively narrow claw that expands broadly in the upper portion of the segments. Duveen, thinking of native canoes, says "paddle-shaped." The petal surface and edges are rippled and tend to fold back along the midribs. The lip seems small by comparison, only 14 mm long, and when flattened is 14 mm across the lateral lobes. The oblique subquadrate lateral lobes have rounded tips and are 8 mm long. The midlobe measures 8 mm across and is transversely oval with a slight notch at the apex. The white and yellow callus terminates in seven to nine roughened veins. The lip is flushed with orchid-magenta, and the veins are darker tones of the same. The column is white with magenta markings and has auricles.

No doubt we will soon hear of selfed seedlings and new hybrids made with this new species. It has a fascinating flower and could introduce new genes for tessellation and petal shape into the gene pool for the cattleyas and their relatives.

Peacock Encyclia

Brazil

(FIGURES 1-3, 2-11; PLATE 30)

Encyclia gallopavina (Reichenbach f.) Porto and Brade. 1935. *Rodriguésia* 1(2):29.

Synonyms

Epidendrum gallopavinum Reichenbach f. 1855. *Bonplandia* 3:219.
Epidendrum purpurachylum B. Rodrigues. 1877. *Gen. & Spec. Orch. Nov.* 1:49.
Encyclia purpurachyla (B. Rodrigues) Porto and Brade. 1935. *Rodriguésia* 1(2): 29.

This rather small-flowered *Encyclia* species has a large name. It is comparatively rare in cultivation but is readily distinctive. I have only seen one plant in Brazil, in 1989 in the collection of an older gentleman affectionately named Papa Noel. I do not know his real name. His plants were collected near Rio de Janeiro. Rodrigues says that they were collected at Queimados and at Parahyba do Sul growing on mimosa and on myrtle trees. They were flowering in August.

The type specimen, Gallo #1694, is conserved in the Reichenbach Herbarium, sheet #46705. A camera lucida drawing of the plant is at Kew and a copy at Ames Herbarium by Ames. The plants are comparatively small with 2 cm tall rounded pseudobulbs and two leaves about 15 cm long. They produce a generous panicle about 60 cm high that has many flowers. The lip has distinctive trapezoid-quadrate lateral lobes, each of which is a little larger than the midlobe, and it has a triangular obtuse tip.

Herbarium notes state that the sepals and petals are pale green slightly tinged with brown. Rodrigues says maroon. The lateral lobes of the lip are purple and turned up at right angles to the axis of the flower in natural position. The midlobe has a distinct purple central spot, and the sides are turned down to be saddle-shaped. Three main veins on the center of the midlobe extend toward the apex from the callus, and the surface of the midlobe is coarsely verrucose. The petals are distinctly spatulate with a

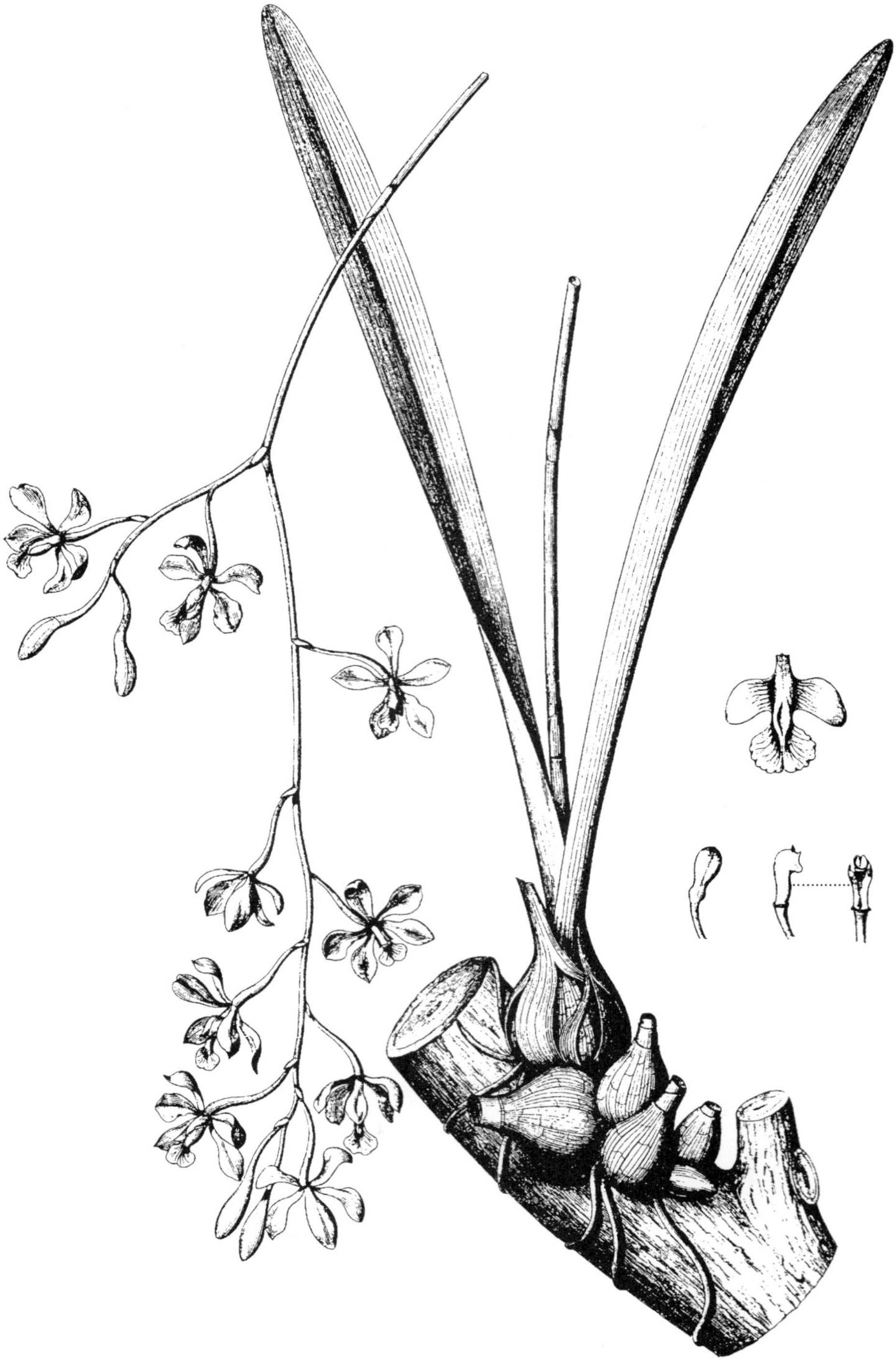

Figure 2-11. *Encyclia gallopavina.* Type drawing by Barbosa Rodrigues published as *Epidendrum purpurachylum* by Cogniaux. Plate 17 of *Martius* 3(5).

narrow claw about one third their length. Rodrigues states the flowers smell of tea throughout the morning.

Measurements from my specimen card show the dorsal sepal 15 × 7 mm; the petals 15 × 8 mm; the lip 13 mm long and 12 mm across the tips of the lateral lobes of the flattened lip and 5 mm across the midlobe with its central magenta spot surrounded by a white rim. A short claw can be found at the lip base and an equally long isthmus between the lateral lobes and the midlobe. The lip shoulders appear "padded" and then drop down to the rounded tips of the lateral lobes. The column is slightly humped, 7 mm long, and the pedicellate ovary 22 mm. The column has small points rather than auricles and is white with purple streaks. The anther cap is "horned" and covered with red speckles. The ovary surface is smooth.

This cute, neat little species would be popular for hobbyists interested in miniatures with large floral displays. Who will be the first to raise it from seed?

Garzón Encyclia

Colombia

(FIGURES 1-3, 2-12; PLATE 31)

Encyclia garzonensis Withner. *Sp. nov.*

Encyclia epiphyticis typicis; plantis notibilis propter sepalis petalisque valde reflexis; panicula longa, laxa, multiflora; labellis valde plicatis; sepalis et petalis dilutis virido-brunneis, textura exigua; labello albis, venis purpurascens, lobo terminali ephippiomorphis; columna albis, auriculatis; ovario glabris. Holotype: Withner, ex hort., without a number, originally collected by George Kennedy near Garzón, Colombia. Deposited in the Oakes Ames Orchid Herbarium (AMES).

For many years I have had a plant in my greenhouse that has defied identification. It was a division of a plant that had been collected in Colombia by George Kennedy. George prided himself, among other things, in having all his plants labeled as to origin and, when possible, properly identified. He had collected most of the plants in his collection himself. On the label of this plant it still says, "Collected on the road to Garzón." I have never been able to match it with any species described for Colombia, and I now have an appropriate moment to describe it as a new species as I write this chapter. Garzón is southeast of Cali on the western side of the Cordillera Oriente.

The flowers are of medium size and thin textured. The sepals and petals are a pale green-tan. The lip is white with purple veins. The column also is white with auricles near its apex. The ovary surface is smooth. There are many flowers widely spaced on a long panicle that is widely branched. The lip is folded down strongly along the midline so that the two halves of the midlobe touch; it gives the impression of a saddle when viewed from the side.

The sepals are 16 × 4 mm, lanceolate-ovate in shape, with an acute tip. The petals are 16 × 4.5 mm, obovate and somewhat clawed, and with acute apices. The lip is 17 mm long, with a 3.5 mm long clawed base, and 17 mm across the lateral lobes. The lateral lobes are nearly at right angles to the axis of the lip when it is flattened, extending 6–7 mm. They are 2 mm long, narrow at their bases and widen toward their rounded tips. The midlobe has a 2 mm isthmus and is rounded, 6 mm wide × 7 mm long, including the rather acuminate apex. The central veins from the forcipate callus extend to the tip of the lip and provide a rib for the folding down of the two sides. The column is slightly humped, with auricles, and a white anther cap. The ovary surface is smooth.

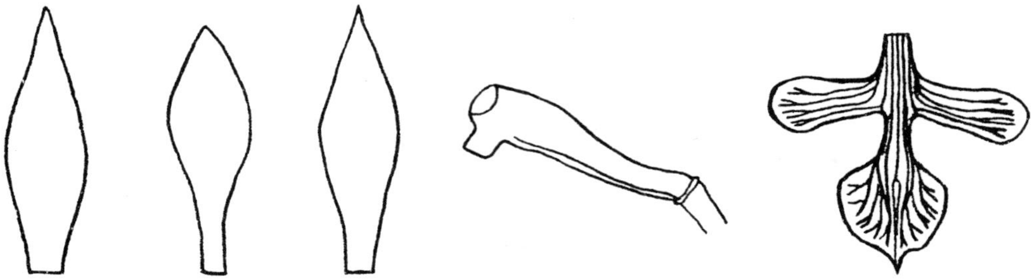

Figure 2-12. *Encyclia garzonensis*, a species found by George Kennedy on the road from Cali to Garzón, Colombia. Please see text for details. It is notable for its distinctly reflexed sepals and petals and the folding down of the midlobe along its center line. Drawing by Michael Ashe.

Gonzalez's Encyclia

Brazil

(FIGURE 1-3; PLATE 32)

Encyclia gonzalezii Menezes. 1991. *Orchid Digest* 55(1):24.

The most distinguishing feature of this species is that the lateral lobes of the lip are nearly diamond-shaped, or rhombic, in technical terms. No other *Encyclia* species has this feature to the same extent. The type specimen was collected in June, 1990, during the Brazilian fall-winter, at 1000 m elevation, by Lou Menezes and Nicola Celso Cazelato. It was growing as an epiphyte in a gallery forest on the border of Brasilia City and Goias state. The type specimen is conserved at the herbarium of the University of Brasilia. The species is named in honor of Father Jose Gonzalez Raposo of Rio Claro City in São Paulo, who is a student of Brazilian orchids.

The type description tells us the pseudobulbs are ovoid-oblong and carry one or two leaves measuring 21 cm long × 1 cm wide. The flower stalk is upright, 35 cm, a raceme with seven flowers that have little fragrance. The lanceolate sepals and petals are green with brownish veins concentrated toward their apices and are about 14–16 mm long × 5 mm wide, the petals a little shorter and narrower than the sepals. The lateral lobes, rhombic in shape, are comparatively short and bluntly pointed. They are lime green with distinct veining toward their bases. The lip midlobe is white shaded with yellow and has purple veining broken into spots and short lines. The main callus ridges are white marked in purple. The lip sizes are not given. The column is 7 mm long and appears white. No mention is made of auricles or wings on the column, the surface of the ovary, or the color of the anther cap.

Goias Encyclia

Brazil

(FIGURE 1-3; PLATE 33)

Encyclia goyazensis Menezes *ex* Fowlie. 1991. *Orchid Digest* 55(2):94.

This new species was found and described by Lou Menezes in an article by Fowlie. About all that is known of it is the description in the article and the photograph that accompanies it. The epiphytic plants came from the region of Marzagão and Caldas Novas and grow at about 600 m in a gallery forest. They were flowering in October and November, the Brazilian spring. The sepals and petals are greenish brown, and the lip is white with a pinkish flush in the callus area particularly over the veins. From the color photo, the flowers have a yellowish cast around the midlobe margins. The lateral lobes are distinctive in having a well-shouldered triangular tapering shape coming to a point. They are green-brown veined at the shouldered base and white with a yellow flush for the balance of their length. On a flattened lip they have a winglike appearance with the rosy-veined, fleshy, white callus ridges between.

The measurements in the description indicate pseudobulbs 8–9 cm tall, two to three leaves 60 cm long × 3 cm wide, and a panicle of 20–25 flowers to 110 cm high. The dorsal sepal is 6 × 20 mm; the petals about the same. Both sepals and petals are oblanceolate-elliptic in shape, the petals a bit narrower. The lip midlobe has a short isthmus and the callus with three larger veins emerging over the base of the disc. The description gives no lip measurements. The column with a greenish base and whiter apex with yellow and rosy flushes is 11 mm long and appears to have wingy points instead of auricles, which are not mentioned in the description. The flowers are slightly fragrant. We need to know more from cultivation about these large plants.

Granite-inhabiting Encyclia

Brazil, Guyana, Suriname, Venezuela

(FIGURES 1-3, 2-13; PLATE 34)

Encyclia granitica (Lindley) Schlechter. 1919. *Repert. Spec. Nov. Regni Veg.*, Beihefte 6:74.

Synonyms

Epidendrum graniticum Lindley. 1841. *Hooker Journ. Bot.* 3:83.

Epidendrum spectabile Focke. 1853. *Bot. Zeit.* 11:229. (Probably = *Enc. granitica*, not *Enc. oncidioides.*)

Misapplications (*auctor non*)

Encyclia patens (Hooker) Rafinesque = *Encyclia patens* Hooker (please see). (There was no valid *Epidendrum patens* first.)

Epidendrum oncidioides Lindley. 1833. *Bot. Reg.* 19, t. 1623 = *Enc. oncidioides* (Lindley) Schlechter. 1914. *Die Orchideen,* ed. 1, p. 210.
Epidendrum oncidioides Lindley. 1833. *Bot. Reg.* 19, t.1623 = *Encyclia granitica* (Lindley) Schlechter. 1919. *Repert. Spec. Nov. Regni Veg.,* Beihefte 6:74. Note these two species were often mistaken for one another.
Epidendrum selligerum = *Encyclia granitica.* (Please see *Enc. selligera* in Vol. V.)
Encyclia lunantha Schlechter. 1919. *Repert. Spec. Nov. Regni Veg.,* Beihefte 6: 40. Possibly a mistaken spelling of *leucantha*?

This species had a poor beginning, as it was always being mixed up with *Encyclia oncidioides.* The rupiculous plants were found growing on granite outcrops and ridges by Richard Schomburgk along the river Corentyne in Venezuela, also in similar places along the rivers Cayuni and Guiana "among boulders where a little soil has collected." Schomburgk described the flowers as green dotted with purple, the lip white with a

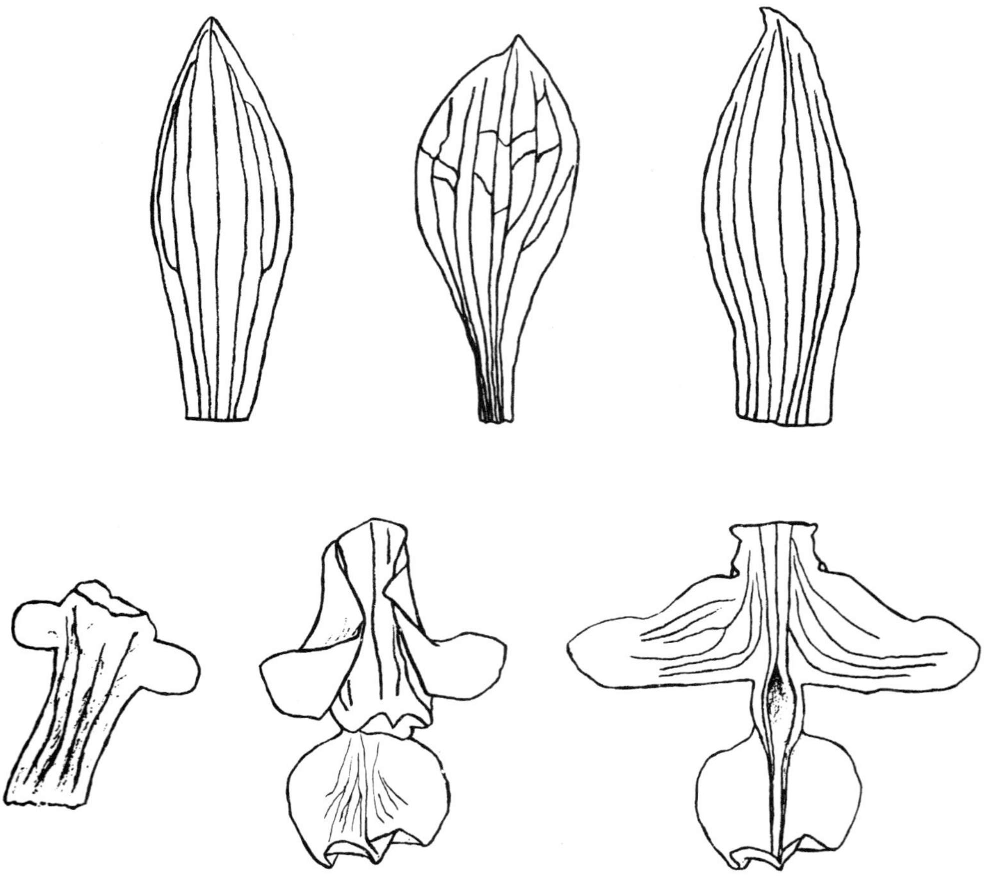

Figure 2-13. *Encyclia granitica.* A flower from the type specimen, Schomburgk #195, conserved in the Lindley Herbarium at Kew Gardens. This species has often been confused with *Enc. oncidioides. Top row, left to right:* dorsal sepal, petal, and lateral sepal. *Bottom row, left to right:* column from above, lip and column, lip. Camera lucida drawing by Antonio Toscano do Brito.

purple stain at its base, the flowers aromatic, and the flower stalk 1.8 m high. Lindley, naming the Schomburgk #195 type specimen, which today is conserved at Kew, wrote that he had only seen portions of the flower panicle. The plants have somewhat elongated rhizomes between growths, 3 cm according to Dunsterville and Garay (1961, pl. 455, p. 151), who also note that they are covered with whitish dried bracts. The growths are generally two-leaved, 30 × 2 cm, considerably smaller than the larger plants of *Enc. oncidioides* that often have three leaves per growth. Pseudobulbs tend to be thin and narrow, 5–6 cm long. The plants flower in the summer.

The sepals are green, and I have always found that their deep maroon spotting ("puce" according to Dunsterville) is practically a hallmark for the species, especially since the sepals as well as the petals have a narrow yellow margin to set them off. The petals are more yellow-green and are also spotted. The lip varies from whitish to creamy color as it ages, and it has a small fan of purple marks on the disc and the basal parts of the lateral lobes. Although the flowers of *Encyclia oncidioides* may have a blotchy coloring, they are brownish yellow, not greenish and puce. The lateral lobes of *Enc. granitica,* when the lip is flattened, are oblong with rounded tips and point out and down. The midlobe is rounded and about 6 mm across with a crenate or ruffled edge. It has a distinct point, and the callus produces a central rib that goes to the apex. The column is green with purple markings, has distinct auricles and a broad apex, and the ovary surface is smooth. The anther cap is yellow. The sepals are oblanceolate, pointed, and have a narrowed base, measuring 16 × 6 mm on my specimen. The petals are spatulate, pointed, and clawed in the lower portion and measure 19 × 7 mm at their widest. The lip is 18 mm long and 18 mm across the lateral lobes; the isthmus is 2.5 mm.

A specimen plant exhibited by M.G.R. Orchids in Florida was granted a CCM award by the American Orchid Society. It had 616 flowers on 22 inflorescences, growing on a tree fern slab. The natural spread of individual flowers was 31 mm.

Guiana Encyclia

Guyana, Suriname, Venezuela

(FIGURES 1-4, 2-14; PLATE 35)

Encyclia guianensis Carnevali and Romero. 1994. *Lindleyana* 9(1):59–70.

Synonym

Epidendrum affine Focke. 1853. *Bot. Zeit.* 11:341.

Misapplications (*auctor non*)

Epidendrum affine Reichenbach f. 1856. *Bonplandia* 4:327 = *Epidendrum cnemidophorum* Lindley.

Encyclia affinis Schlechter, *nom. nov.* 1918. *Beih. Bot. Centralbl.* 36(2):471 = *Epidendrum cnemidophorum* Lindley.

Epidendrum affine Richard. 1850. In Sagra's *Hist. Cuba* 11:235, t. 78 = *Epidendrum fucatum* Lindley.

When describing some orchids from Suriname in 1853, Focke described *Epidendrum affine* as a new species, recognizing that it belonged to Lindley's section *Encyclia* of that genus. We cannot, however, now make the transfer to the genus *Encyclia,* since that epithet is already preoccupied by an *Enc. affinis* published by Schlechter in 1918. Schlechter based his name on *Epi. affine,* a Reichenbach name that was already illegal as there were two previous *Epi. affine* species epithets published, and by our rules only one example of a published name is permitted within a genus. This example is worse than usual with four homonyms involved. Even though Focke's *Epidendrum* name was for the species we have here, it cannot be transferred to *Encyclia,* and that makes the Carnevali and Romero name the only valid epithet for the species.

Focke notes that this species is not *Encyclia granitica* (synonym *Epidendrum spectabile*), but differs by having top-shaped pseudobulbs, short foliage, a racemose flower stalk, and smaller flowers. The sepals and petals were green, the lip white with the apex revolute, and the column was auricled with the anther cap yellow. The sepals and petals were lanceolate-spatulate and acute. The lateral lobes of the lip were oblong and rounded. The midlobe was suborbicular and pointed.

I believe that this description matches the *Encyclia guianensis* described in 1994 by the Carnevalis and Romero. That plant came from Bolívar state in Venezuela along the side of the road to the south of Dorado at 500 m elevation. The plants have pyriform pseudobulbs 1–4 cm tall and usually two broad, short, obtuse leaves to 22 cm long × 2.3 cm wide. The flower stalks are upright, densely flowered panicles with two to four branches. The type specimen for the species is Steyermark, Dunsterville, and Dunsterville #93048 conserved in Venezuela with an isotype specimen at the Ames Herbarium. The drawings are Dunsterville #852 and #1415. The latter drawing is not identified in the article by Carnevali and Romero. The flowers have a natural spread of 2.0–2.5 cm.

The sepals and petals are light greenish yellow or may have maroon flushes with noticeable veining toward their apices. The lip is white or creamy with three reddish veins on either side of the three thickened central veins. The lateral lobes have greenish bases veined with purple and purple-pink apices.

Measurements of the flower parts are as follows: dorsal sepal 10–12 × 3.3 mm; petals about the same, a little wider toward their apices and more clawed than the sepals toward the base. Both sepals and petals are obovate. The lip midlobe is 8.5 mm long × 8 mm wide and is spade-shaped with a pointed apex. The 8 mm column has distinct auricles. The ovary surface is not mentioned, but the flower rachis is verruculose. An interesting note about the species's habitat is that it grows in the same area as *Encyclia chloroleuca.* Also, the authors had earlier (1988) identified this plant as *Enc. picta* from Guyana, but on reconsidering the flowers they decided it was a new species. *Encyclia picta* has a narrower, more elliptical midlobe shape, and it lacks auricles on its column—one of the clearcut differences.

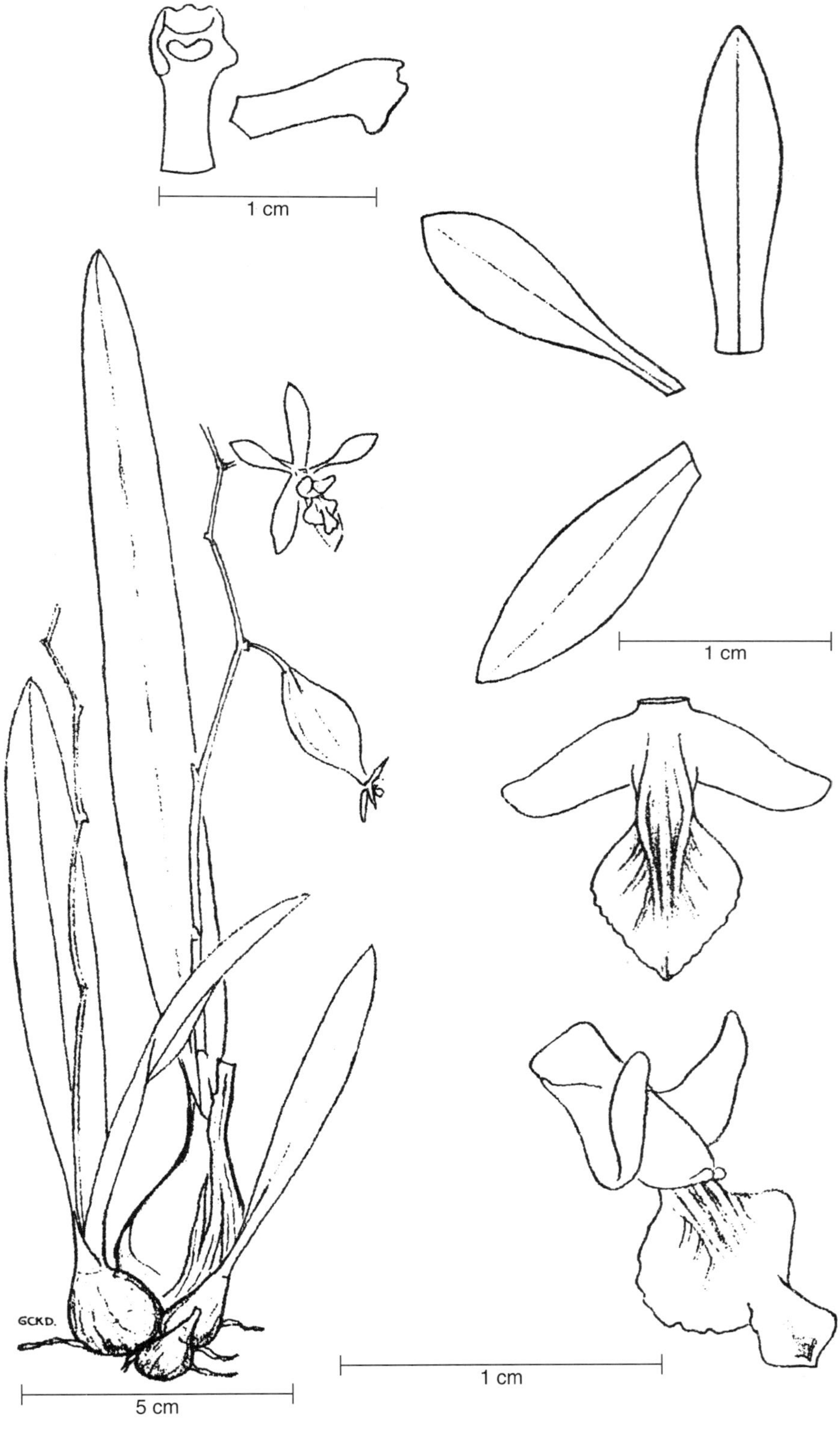

Figure 2-14. *Encyclia guianensis.* Drawing #1415 by G. C. K. Dunsterville.

Violet-scented Encyclia

Brazil, French Guiana, Guyana

(FIGURES 1-4, 2-15; PLATE 36)

Encyclia ionosma (Lindley) Schlechter. 1914. *Die Orchideen,* ed. 1, p. 209. Also see Lindley *ex* Stein, 1892, *Orchideenbuch,* p. 232.

Synonyms

Epidendrum ionosmum Lindley. 1838. *Bot. Register* 24: 49, misc. 87.
Epidendrum euosmum Reichenbach f. 1876–1877. *Linnaea* 41:79.
Encyclia flabellifera Hoehne and Schlechter. 1926. *Arq. Bot. Est. São Paulo* 1: 246, t. 6, fig. 1.
Encyclia euosma (Reichenbach f.) Porto and Brade. 1935. *Rodriguésia* 1(2):29.

Misapplication (*auctor non*)

Epidendrum flabellatum Lindley. 1853. *Fol. Orch. Epid.,* p. 6 = *Epidendrum candollei,* México.

The problems tracking down this species all fell into place by accident when I asked Antonio Toscano de Brito to look for the type specimen of *Encyclia flabellifera* at Kew. He found it, in the Lindley Herbarium, a single flower from a plant imported by Loddiges from Essequibo. Toscano was a graduate student at the time working on his Ph.D. degree at Reading University in England, but doing his research on orchids at Kew. He knew Brazilian orchid species well, having been Pabst's research assistant for several years. A full-page color plate of *Enc. flabellifera* is in Hoehne's *Iconografia* (1949, t. 112). I could never get it to match any of the specimens I had seen at Kew or elsewhere, and I did not quite know what to do about it, especially since the epithet *flabellifera* was close to, if not homonymic with, *Enc. flabellata* from México.

After finding the type specimen, Toscano dissected the flower and then made a camera lucida drawing of it (Figure 2-15). The long, narrow lateral lobes of the lip curving down with parallel sides and crenate apices were distinctive, as was the rough-edged fanlike midlobe. Besides that, notes showed that the fresh flowers had the strong scent of violets. The synonyms, more or less, began then to fall into place, and after more research Lindley's *Epidendrum ionosmum* finally provided the basionym for the proper name of this species.

The type specimen for *Epidendrum ionosmum* was a plant collected by Thomas Spencer in Guyana and sent from Demerara (now Guyana) to Loddiges where it was cultivated. The Reichenbach type for *Epi. euosmum,* a similar flower and somewhat of a mystery, was collected in Brazil by Henri Blunt, a collector for Hugh Low and Company in England, and grown by that company in 1863. It is in Vienna, #46892. An outline sketch on the herbarium sheet shows the flower details, the narrow parallel-sided elongated lateral lobes, the verrucosities of the veins on the fanlike shape of the midlobe, and the serrate and crisped margin. John Day (1884, 42:23) painted this specimen for his records, writing,

> It is not a handsome species, but the scent is exquisite and filling the house in which it is growing (at Hugh Low and Company). The insignificant and dull colored orchids frequently smell much more powerfully than the large and brilliantly colored ones, many of which are quite scentless—so no doubt the insects that fertilize them are attracted by scent.

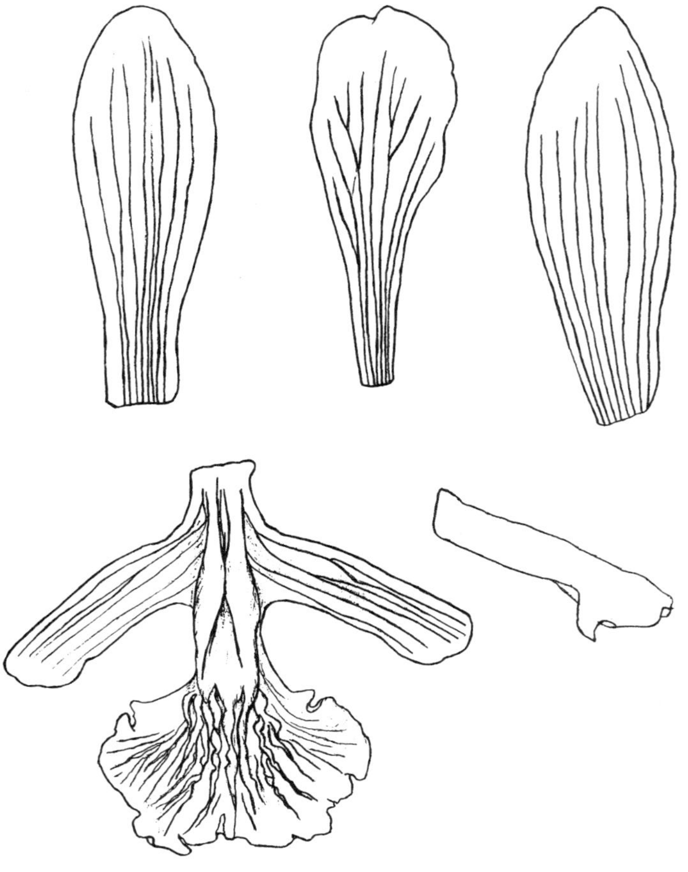

Figure 2-15. *Encyclia ionosma.* A flower from the type specimen of *Encyclia flabellata,* #936, a synonym. The specimen is conserved in the Pabst collection in Brazil. *Top row, left to right:* dorsal sepal, petal, and lateral sepal. *Bottom row, left to right:* lip (note the long, narrow lateral lobes), and side view of column. Camera lucida drawing by Antonio Toscano do Brito.

The fragrance of violets distinguished this epidendrum and gave it the name of "truly fragrant," as its Latin epithets of *ionosma* or *euosma* would be translated. It was likened to *Encyclia alata* from Central America when it was first described, mostly, I believe, because of the lip color and verrucosities. The identity of *ionosma* has always been debated, usually because the plants had been called *Encyclia flabellifera.*

Day (35:41) also sketched *Epidendrum ionosmum* and (10:15) painted a single flower and a plant of *euosmum.* With all the specimens, descriptive information about this species is not lacking. It is comforting, for a change, to have good documentation for the proper name. That is not the situation for other encyclias.

The plants have two or three leathery leaves that are erect and dark green to 25 cm long × 3 cm wide. The plants are comparatively short with clustered small, round pseudobulbs that are shiny and may reach 4–5 cm long. The green flower stalk is a little taller than the leaves, sometimes is branched, has comparatively few flowers, and usually is roughened to the touch. The plants are summer flowering in cultivation and make good specimens—especially when there are several flower stalks with their fragrant blooms. They grow best on cork or in a basket mounted in a bright location. The plants have often been misidentified in cultivation—everything from *Encyclia odoratissima* to *Enc. tampensis.*

Sepals and petals are green-brown with a yellow border, obovate and slightly concave, to 17 × 6 mm. The lip is about 16–17 mm long, pale yellow with reddish veining broken into many varicosities over the disc. The midlobe is variously described as fanlike in shape or transversely oblong, with a crisped or crenate margin, and slightly indented at the apex. It can spread to 10 mm wide and be only 7 mm from isthmus to tip. On a flattened lip the lateral lobes extend, almost winglike, with their widely curving and parallel sides, 8–9 mm long with their ruffled apices that may measure 15–17 mm apart. The callus keels are white and situated mostly on a narrow, elongated isthmus 3–4 mm long. The 1 cm humped column is yellow-green and has auricles, and the ovary surface is smooth. The anther cap is yellow.

This species could be confused with *Encyclia angustiloba* from Ecuador, Peru, and Bolivia, except that it has a slightly larger lip with more exaggerated lateral lobes, subspatulate petals not greatly narrowed at the base, and the ovary surface is smooth, not asperous.

Ivón's Encyclia

Brazil?, Venezuela

(FIGURES 1-4, 2-16)

Encyclia ivonae Carnevali and Romero. 1994. *Lindleyana* 9(1):65.

This species is described by Carnevali and Romero and named after Carnevali's wife, Ivón Ramírez. They have stated that it is not *Encyclia recurvata* nor *Epidendrum halatum* (two names for the same species), but is more like *Enc. diurna,* but smaller. At this point I am a little confused about its true nature. It is not pictured in their article, but it is well described. The type specimen is Steyermark #60870 at the Ames Herbarium, and it was collected in the extreme southeastern state of Bolívar on the lower southeastern slopes of Carrao-tepui at 1,675–19,870 m. That area is on the frontier with Brazil, so Brazil is also included in the probable distribution of the species. It was flowering in December when collected and was found growing epiphytically, occasionally on rocks, in gallery forests in well-exposed locations, especially where forest and savanna habitats met.

The plants can reach 150 cm tall including the flowers, and they are often flushed with purple from exposure to the light. The pseudobulbs are clustered, reaching 12 cm tall, more or less pyriform, and two- to three-leaved. The leaves are to 55 cm long × 5 cm wide. Paniculate flower stalks may be 100 cm long and erect, and the lower branches may rebranch. The fleshy flowers are strongly fragrant and may last in good condition for 3–5 weeks. Although the flowers may vary in color from green with maroon tinges to dark brown-purple with green toward their bases, the sepals and

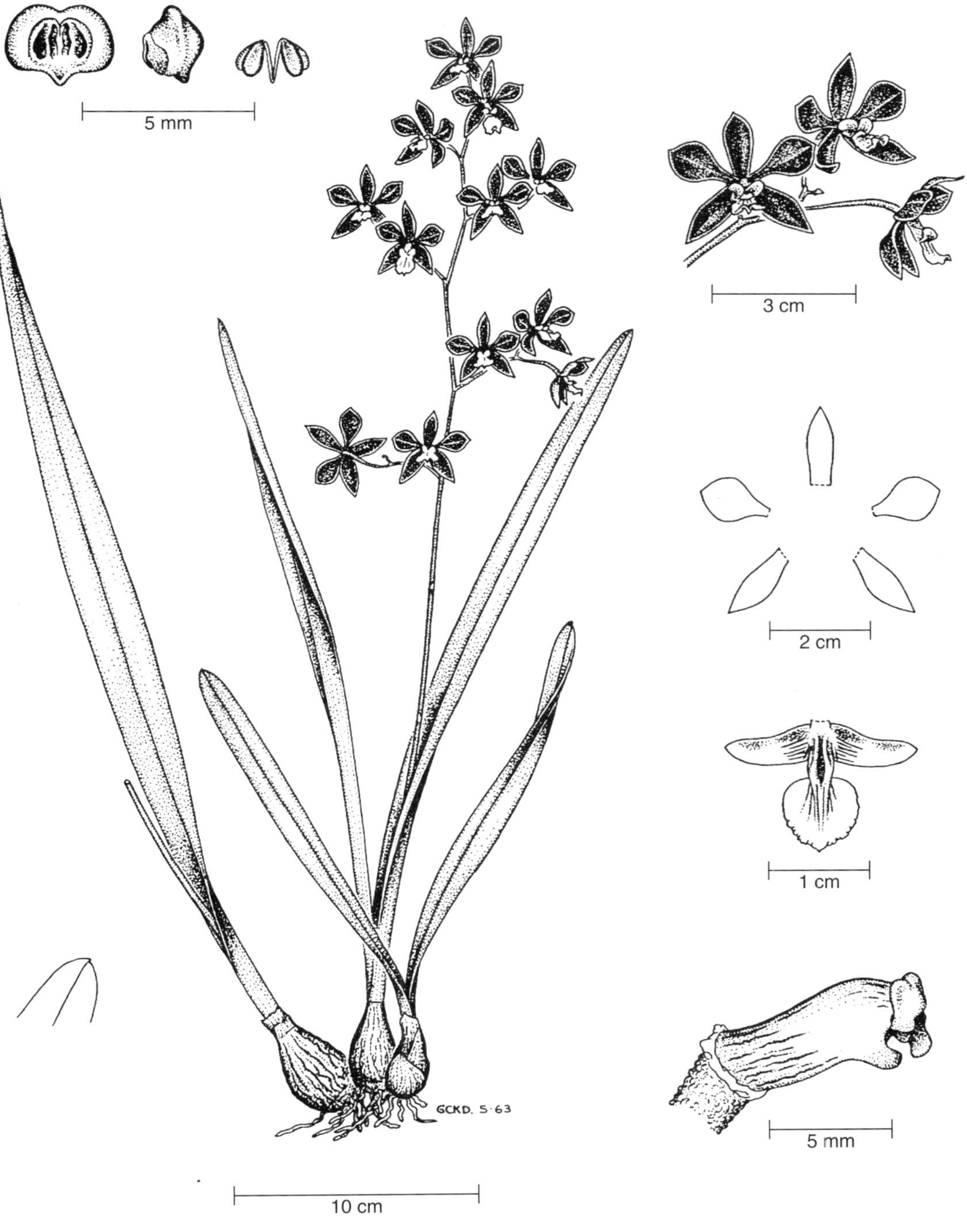

Figure 2-16. *Encyclia ivonae.* Drawing #772 by G. C. K. Dunsterville. It was originally identified as *Epidendrum halatum* on p. 233 of Dunsterville and Garay's *Field Guide* (1979).

petals have distinguishing clear margins. The lips are creamy white or yellow with purple veining, particularly the central rib. Columns are green at the base then yellowish toward the apex and with flushes of maroon or brown. They are distinctly auricled. Anther caps are bright yellow.

The sepals are elliptic and pointed, to 16 × 6 mm; the petals are obovate and widely spatulate in shape with obtuse or almost rounded tips to 16 × 9.5 mm; labellum is to 16 mm long, spreading to 18 mm across the lateral lobes on a flattened lip; the midlobe is 8.5 long × 8 mm wide. Three thickened central veins extend to the apex of the midlobe. The lateral lobes are 7.5 mm long × 3 mm wide at their bases and are narrowly triangular-ovate with rounded apices.

What these authors are really saying in their article is that *Encyclia ivonae* is a new species being described, and that the illustration in Dunsterville and Garay (1959–1976, 6:127) labeled *Epidendrum halatum* is not *halata* nor *recurvata,* but is the new species *ivonae.* We are now left with trying to find an illustration of the true *Enc. recurvata* (please see) or *halata. Encyclia ivonae* has not been in general cultivation though mention is made in the article that it may be growing in some Venezuelan collections. Also see *Enc. diurna,* another species with which it has been confused.

Broad-petaled Encyclia

Brazil, Guyana, Suriname, Venezuela

(FIGURES 1-4, 2-17)

Encyclia latipetala (Schweinfurth) Pabst. 1967. *Orquídea* (Brazil) 29(1):9.

Synonym

Epidendrum latipetalum Schweinfurth. 1948. *Bull. Torrey Bot. Club* 75:219, fig. 16.

This plant with stubby, upright, wide leathery leaves and only a few large green flowers in a short raceme has rarely been collected and is not well known. The species would seem to be close to *Encyclia remotiflora,* but as Carnevali and Ramírez (1988) point out, the flowers of *Enc. latipetala,* as well as those of *Enc. pachyantha,* turn black on drying, while those of *Enc. remotiflora* turn brown. The few specimens to study in a case like this one often make distinguishing the species difficult. Insufficient specimens prevent study to decide between species' differences and just variation. The type specimen becomes a critical factor in such decisions. The three species have been mixed up because of their general similarities.

The *Encyclia latipetala* flower was from a plant collected on the Kaieteur Plateau, Guyana, by McGuire and Fanshawe #23327, conserved at the New York Botanical Garden Herbarium. The plants are described as growing in the shade, being "terrestrial," and the two leaves, to 16 cm long × 2.7 cm wide, seem to be of unequal size—a unique and tell-tale character. The pseudobulbs are 2.5 cm long.

Flower measurements and shapes include sepals 23 × 9 mm, oblanceolate and pointed; petals obovate-spatulate, to 22 × 11–12 mm. The lip is 18 mm long and deeply lobed, and has oblongish lateral lobes with widened tips that have slightly crenate margins. The midlobe is ovate-rounded at the

Misapplications (auctor non)

Epidendrum flavum Mutel in *Mem. Soc. Sci. Lille* 151, fig. 1, 1843 = *Epidendrum mutelianum* Cogniaux.

Epidendrum flavum var. *fuscosepalum* Hoehne. 1912. *Comm. Linos Teleg. Matto Grosso. Annex* 5, Bot. 4:19 = unknown species.

This species has been considered synonymous with *Encyclia patens* or *Enc. conchaechilum.* A drawing by Toscano (Figure 2-10) from the Gardner #3456 type specimen of 1841 conserved at Kew shows the details of its flower structure. The plant came from "near the Caza Pintada in decaying vegetable matter" in São Paulo, Brazil. The column is narrow and straight and shows only slight wings or widenings toward the apex and no auricles as such. These characteristics eliminate it from consideration as synonymous with *Enc. patens,* which has definite quadrate auricles. Also *Enc. flava* has long narrow leaves, not comparatively short wide ones like *Enc. patens,* but it does have a paniculate inflorescence, and is rather densely flowered at about leaf height, characters that are typical of *patens.* The lip details of *Enc. flava* show five raised veins running out from the callus, not three as in *Enc. patens,* and the lateral lobes of the two species are also different in their

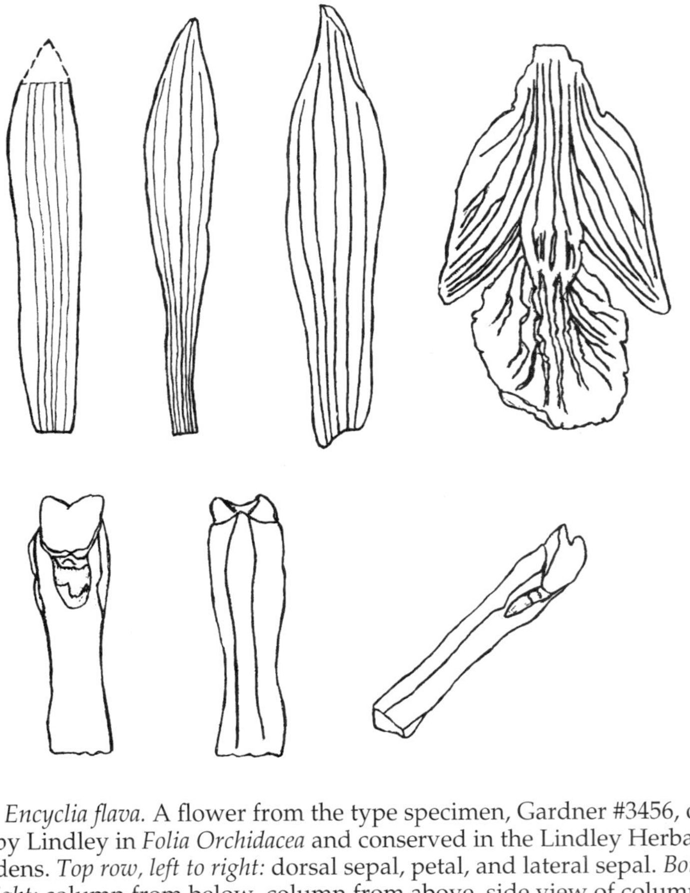

Figure 2-10. *Encyclia flava.* A flower from the type specimen, Gardner #3456, of 1841, mentioned by Lindley in *Folia Orchidacea* and conserved in the Lindley Herbarium at Kew Gardens. *Top row, left to right:* dorsal sepal, petal, and lateral sepal. *Bottom row, left to right:* column from below, column from above, side view of column. Camera lucida drawing by Antonio Toscano do Brito.

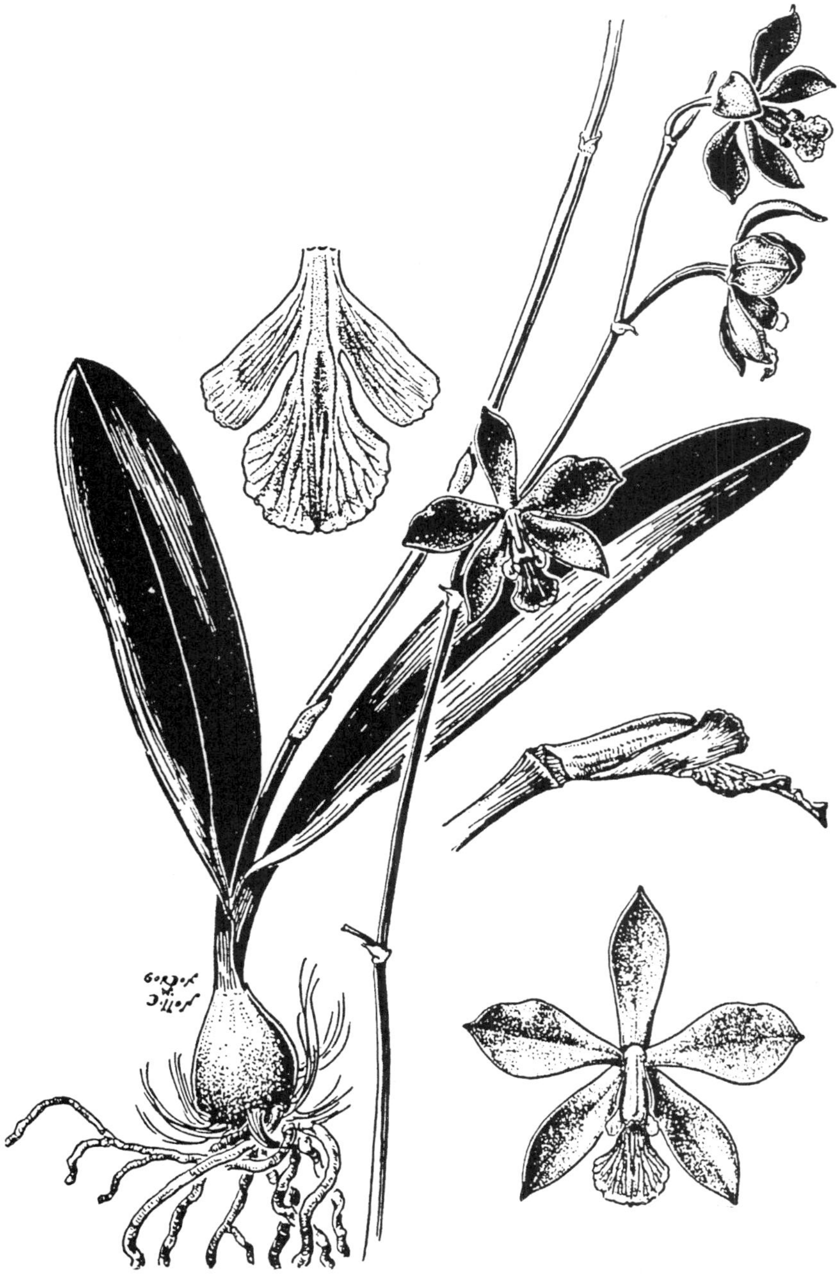

Figure 2-17. *Encyclia latipetala.* Drawing by Gordon W. Dillon. With permission of the Torrey Botanical Club.

tip with wavy margins, and a blunt or even retuse apex, sometimes described as "abruptly acute." The midlobe arises from a wedge-shaped base with tissue on either side of the callus that is recurved. The column is curved, auricled, and about 12 mm long; the ovary surface is smooth. These features are well illustrated in the type illustration by Gordon Dillon from the *Bulletin of the Torrey Botanical Club,* reproduced here as Figure 2-17.

The plants flower from late spring through early summer. Size of the plant would seem to eliminate it from consideration as a synonym of *Encyclia pachyantha* or *Enc. remotiflora,* also the wedge-shaped base of the midlobe of the lip, the extra-wide petals, the crenate tips of the lateral lobes of the lip, and the unequal size of the leaves. Drawing #180 by Dunsterville (1979) is equally detailed.

White-flowered Encyclia

Colombia, Venezuela

(FIGURE 1-4; PLATE 37)

Encyclia leucantha Schlechter. 1919. *Repert. Spec. Nov. Regni Veg.,* Beihefte 6:40.

Synonym

Epidendrum leucanthum (Schlechter) Schnee. 1945. Drawing made in Berlin under Schlechter's supervision, the lectotype copy of which is AMES #26791 printed in Pittier, 1945, *Cat. Flor. Venezuela* 1:215 (reported in Carnevali et al., 1994).

The type specimen for this species was sent to Schlechter in 1905 for identification. It had come from Venezuela and had flowered in the collection of K. W. John from Andernach on the Rhine in western Germany. Schlechter noted especially the hairy callus pad on the lip and suggested that the species was otherwise close to *Encyclia stellata,* a statement that I do not understand today, even though *Enc. stellata* may have a few hairs along the callus ridges. A photograph of the original inflorescence with detailed flower and column sketches, and a drawing of a leaf, are at Harvard and provide an iconotype record for the species. The flowers were described as a creamy white, though Schlechter comments in his write-up on the species that he was away from home when it arrived, and it was not in the best condition when he examined it. The short downward-pointing wide and rounded lateral lobes of the flattened lip with the hairy callus between are distinctive. Add to this the scabrous ovary and the verrucose veining on the lip midlobe, the three central veins especially thickened, and the apical wings of the column that fold in as though they were auricles, and the species is well characterized. The lip is 13 mm long. One specimen, Wurdack and Monachino #41202, collected in January, 1956, in Bolívar state of Venezuela on the Cerro Baraguan, has some flower color notes. The sepals and petals were dull yellow, the lip white becoming yellow, streaked with purple. This specimen is conserved at the Ames Herbarium. I might also add that the anther cap is dark red and bifid in appearance, making this species belong in what might be called the Amicta Group or Complex (please see *Enc. amicta* above). It has no auricles on the column. Drawing #180 by Dunsterville (1979) portrays this species well.

Lutzenberg's Encyclia

Brazil, Venezuela

(FIGURE 1-4; PLATES 38, 39)

Encyclia lutzenbergeri Menezes. 1990. *Bol. C.A.O.B.* 3(1):5. Republished in *Orchid Digest* 55(1):25, 1991.

José Lutzenberg, the Environmental Secretary of Brazil, was honored by the naming of this new *Encyclia* species found in Brasilia by Jesus G. Miranda, a member of the local orchid society. It was subsequently found also in the neighboring Goias state, flowering at the end of September. Having the same sort of fragrance, it was considered close to *Enc. osmantha,* but the plants were considerably smaller, had flowers with longer lateral lobes with truncate rosy apices, and the midlobe was tinged with pink and had white edges. The column has auricles, if I interpret the type drawing correctly. The squared off tips on the lateral lobes are a hallmark for the species.

The pseudobulbs were bifoliate, and the leathery leaves measured 19 cm long × 2.3 cm wide with pointed tips. The erect paniculate inflorescence reached 60 cm. The sepals were green with veining, 19 × 5 mm, and acute. The petals were spatulate and veined like the sepals, 18 × 6 mm. The lip midlobe was rounded and had three main purple veins on the disc. Both sepals and petals had a yellowish margin.

Since that time Menezes has written an article (1992) on the variations in *Encyclia lutzenbergeri* and its close relation to *Enc. osmantha* and has named a geographic race as variety *major.* These plants are rupiculous, growing on rock outcrops near Vitoria da Conquista in the extreme southwest of Bahia state. They grow at about 950 m and flowered in Brasilia in October. Further experience with this species and its variety will give us a better understanding of how they fit into the *Encyclia* scheme of things.

Maderoi Encyclia

Colombia, Venezuela?

(FIGURES 1-4, 2-18; PLATE 40)

Encyclia maderoi Schlechter. 1920. *Repert. Spec. Nov. Regni Veg.,* Beihefte 7:155.

Synonym

Epidendrum bacigalupianum Hawkes. 1956. *Orquídea* 18:174.

Misapplication (*auctor non*)

Epidendrum maderoi Schlechter. 1920. *Repert. Spec. Nov. Regni Veg.,* Beihefte 7:138. Also see Mansfeld, 1929, *Atlas der Kordillerenstaaten,* t. 55, no. 212.

Several Colombian encyclias are poorly known today, and they have seemingly been lost among the greater number of showy species, especially cattleyas, from that country. This species is one of them, and we are not certain about its identity. Although it was described and pictured by Schlechter, it has few additional specimens available for study. In addition, Dunsterville sent me a copy of one of his drawings, #1382, that was not identified, and I believe that it portrays this species in Venezuela. The type species for Schlechter's description was from Antioquia, Colombia, at about 1000 m by M. Madero. The collection data with the Dunsterville specimen says "Edo. Táchira Hato Río Frío," a mountainous location near the Colombian border in western Venezuela.

The type description says the plants are 1 m tall with clumped pseudobulbs about 6 cm tall, obliquely ovoid with two leaves. The erect and spreading leaves measure 40–45 cm long × 1.5–2.0 cm wide. The lax paniculate inflorescence has a stalk of 30–60 cm with spreading flexuous branches to 25 cm. Schlechter remarks that the flowers are most like those of *Encyclia oncidioides,* but the plants have less dense panicles, the petals are much smaller, and the rather long and broad lateral lobes of the lip help distinguish it. Nothing is written about the color of the flower parts; only shapes and sizes are mentioned.

The sepals are spreading, oblong, subacute, narrower toward the base, and 19 mm long. The petals are elliptic-unguiculate, subacute, slightly wider than the sepals above the middle with a narrow base, also 19 mm long. The distance between the lateral lobe tips is 2 cm. The callus extends as three main veins onto the midlobe. The lateral lobes are oblong-strap-shaped, obtuse, their apices a little dilated, 9 mm long and 5 mm wide below the tips. The midlobe has a short isthmus of 2 mm, is quadrate-orbicular, apiculate, and slightly undulate, about 11 mm long by 10 mm wide. The semi-terete column is 10 mm long with incurved semi-oblong auricles. The pedicellate ovary is smooth, 25 mm long.

As do many of these *Encyclia* species, the plants grow well on cork slabs in the greenhouse with good light and air, paying attention to giving less water during the winter months when the plants are dormant. This stately and exciting species should be used in hybridizing for its vigor, size, floriferousness, and color.

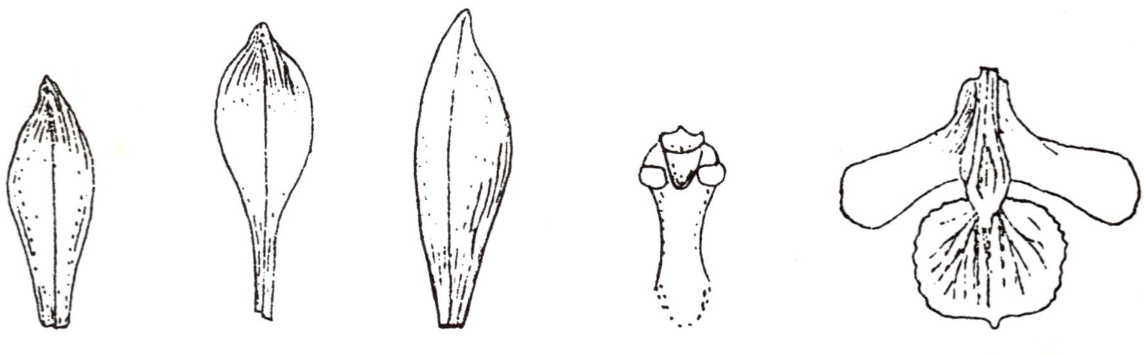

Figure 2-18. *Encyclia maderoi.* This *Encyclia* species from western Venezuela and Colombia is not well recognized today in spite of its notable brightly colored flowers—chartreuse-green sepals and petals and lips with bold, rich red-purple veining on a white background. Adapted from drawing of the holotype specimen, destroyed in Berlin, from Mansfeld's *Atlas der Kordillerenstaaten*, 1929.

PLATE 1. *Encyclia advena*. In the author's collection. Photo by Ron Parsons.

PLATE 2. *Encyclia albopurpurea* at Orquidário Binot. Plant from the river Tocantins, Brazil. Photo by Jack Fowlie.

PLATE 3. *Encyclia amicta*. Photo by David G. Hunt.

PLATE 4. *Encyclia andrichii*. Type collection. Photo by Lou Menezes.

PLATE 5. *Encyclia angustiloba*. Ecuadorian plant in the Ragan collection.

PLATE 6. *Encyclia argentinensis* at Orquidário Binot.

PLATE 7. *Encyclia aspera* at the Guayaquil Orchid Show.

PLATE 8. *Encyclia bracteata*. In the author's collection. Photo by Ron Parsons.

PLATE 9. *Encyclia bragancae* at Orquidário Binot. Photo by Jack Fowlie.

PLATE 10. *Encyclia burle-marxii* in Brazil. Photo by Jack Fowlie.

PLATE 11. *Encyclia caximboensis.* Type specimen. Photo by Lou Menezes.

PLATE 12. *Encyclia ceratistes.* Venezuelan plant in the Dunsterville collection.

PLATE 13. *Encyclia chapadensis.* Type specimen. Photo by Lou Menezes.

PLATE 14. *Encyclia chapadensis* in its habitat of Chapada dos Viadeiros growing on lichen-covered rocks in bright sun. Note the natural redness of the pseudobulbs and the yellow-green of the leaves.

PLATE 15. *Encyclia chloroleuca.* plate 3557 from *Curtis's Botanical Magazine,* 1837.

PLATE 16. *Encyclia conchaechila* at Orquidário Binot. Photo by Jack Fowlie.

PLATE 17. *Encyclia confusa* in Espírito Santo state, Brazil. Photo by Lou Menezes.

PLATE 18. *Encyclia cordigera*. In the Marsh collection. Photo by Trudi Marsh.

PLATE 19. *Encyclia cyperifolia*. Plant collected in Peru by Fred Fuchs.

PLATE 20. *Encyclia dichroma* in Bahia state, Brazil. Photo by Robert Liddell.

PLATE 21. *Encyclia dichroma* habitat on rock outcroppings in Bahia state, Brazil. Photo by Robert Liddell.

PLATE 22. *Encyclia diurna*. Venezuelan plant.

PLATE 23. *Encyclia doeringii*. Painting from Hoehne's *Iconografia de Orchidáceas do Brasil*.

PLATE 24. *Encyclia dutrai* at the Brazilian Orchid Show, 1981.

PLATE 25. *Encyclia duveenii* at Orquidário Binot. Photo by Trudi Marsh.

PLATE 26. *Encyclia edithiana* in Brazil. Photo by Lou Menezes.

PLATE 27. *Encyclia flava* in Brazil. Photo by Jack Fowlie.

PLATE 28. *Encyclia fowliei*. In the Hunt collection. Photo by David G. Hunt.

PLATE 29. *Encyclia fowliei*. Note the pseudobulb shape and the color of the plant.

PLATE 30. *Encyclia gallopavina* at Orquidário Binot. Photo by Jack Fowlie.

PLATE 31. *Encyclia garzonensis.* George Kennedy plant from Garzón area of Colombia.

PLATE 32. *Encyclia gonzalezii.* Type specimen. Photo by Lou Menezes.

PLATE 33. *Encyclia goyazensis.* Type specimen. Photo by Lou Menezes.

PLATE 34. *Encyclia granitica.* Venezuelan plant in the Dunsterville collection.

PLATE 35. *Encyclia guianensis.* Adapted from a Dunsterville photo in Venezuela.

PLATE 36. *Encyclia ionosma* in Orgão Mountains, Rio de Janeiro, Brazil. Photo by Trudi Marsh.

PLATE 37. *Encyclia leucantha.* In the Hunt collection. Photo by David G. Hunt.

PLATE 38. *Encyclia lutzenbergeri.* Adapted from photo by Lou Menezes in the *Orchid Digest.* With permission of the *Orchid Digest.*

PLATE 39. *Encyclia lutzenbergeri* var. *majus*. In the Tom Biggert collection.

PLATE 40. *Encyclia maderoi*. In the Hunt collection. Photo by David G. Hunt.

PLATE 41. *Encyclia magdalenae a*t Quito Orchid Show. Plant from Andrea Niessen collection of Orquideas del Valle, Cali, Colombia.

PLATE 42. *Encyclia megalantha* at Orquidário Binot.

PLATE 43. *Encyclia meneziana*. In the Severin collection, Cupertino, California. Photo by Ron Parsons.

PLATE 44. *Encyclia microtos*.

PLATE 45. *Encyclia microxanthina.* Plants from Paraguay. Photo by Jack Fowlie.

PLATE 46. *Encyclia naranjapatensis.* In the Severin collection, Cupertino, California. Photo by Ron Parsons.

PLATE 47. *Encyclia oncidioides*. Branch of the densely flowered, erect panicle. Plant from the Lineau Robert collection in Brazil.

PLATE 48. *Encyclia oncidioides.* Flower closeup.

PLATE 49. *Encyclia osmantha.* In the Severin collection, Cupertino, California. Photo by Ron Parsons.

PLATE 50. *Encyclia pachyantha.* In the Severin collection, Cupertino, California. Photo by Ron Parsons.

PLATE 51. *Encyclia patens.* In the author's collection. Plant originally from Orquidário Binot. Photo by Ron Parsons.

PLATE 52. *Encyclia pauciflora* in Brazil. Photo by Jack Fowlie.

PLATE 53. *Encyclia pedra-azulensis* in Brazil. Photo by Lou Menezes.

PLATE 54. *Encyclia pilosa.* In the Hunt collection. Photo by David G. Hunt.

PLATE 55. *Encyclia profusa* in Medellín, Colombia.

PLATE 56. *Encyclia quesneliana* in the Serra de Bahiana near Sergipe state, Brazil. Photo by Robert Liddell.

PLATE 57. *Encyclia randiana* at Orquidário Binot. Photo by Trudi Marsh.

PLATE 58. *Encyclia randiana* var. *rondoniensis* in Rondônia, Brazil. Photo by Lou Menezes.

PLATE 59. *Encyclia replicata.* Type painting by Lindley in the orchid herbarium at Kew. Photo by author.

PLATE 60. *Encyclia santos-dumontii.* Photo by Lou Menezes.

PLATE 61. *Encyclia schmidtii* in Mato Grosso state, Brazil. Photo by Lou Menezes

PLATE 62. *Encyclia sclerocladia.* In the Hunt collection. Photo by David G. Hunt.

PLATE 63. *Encyclia seidelii.* In the author's collection. Photo by Ron Parsons.

PLATE 64. *Encyclia spiritusanctensis* in Brazil. Photo by Lou Menezes.

PLATE 65. *Encyclia steinbachii.* In the Mizuta collection in Hawaii.

PLATE 66. *Encyclia tarumana* in Manaus, Brazil. Photo by Lou Menezes.

PLATE 67. *Encyclia thrombodes.* Peruvian plant in the author's collection.

PLATE 68. *Encyclia unaensis* in Brazil. Photo by Jack Fowlie.

PLATE 69. *Encyclia viridiflava* in Brazil. Photo by Lou Menezes.

PLATE 70. *Encyclia viridiflora.* plate 2831 from *Curtis's Botanical Magazine,* 1828. Type illustration for the genus and the species.

PLATE 71. *Encyclia xerophytica* at Orquidário Binot. Photo by Jack Fowlie.

PLATE 72. *Encyclia yauaperyensis*. Plant from Amazonas, Brazil, collected near Iquitos, Peru, by Robert Wilson.

Magdalena Encyclia

Colombia

(FIGURES 1-4, 2-19; PLATE 41)

Encyclia magdalenae Withner. *Sp. nov.*

Pseudobulbis pyriformis, apice ditriphyllis; foliis coriaceis, lineari-ligulatis. Panicula ampla, laxa multiflora formante quasi 1 m; floribus rubro-brunneis. Labellum lobi lateralibus late oblongis, apice rotundatis; lobo terminalo roseo cum venae atrovinosae. Columna apice biauriculata, angulata cum ovario, plus minusve glabra. Holotype: Withner, ex hort., without a number, collected by Withner at Miami International Orchid Show. Deposited in the Oakes Ames Orchid Herbarium (AMES).

I have seen three plants of this species, one at a show in Miami, Florida (presumably from Colombia), one at a show in Medellín, Colombia, and another at a show in Guayaquil. It seems to be a notable and distinctive species, and yet it has eluded me over the years in my attempts to obtain a plant for study or to confirm its identity. I have not been able to match it with any description of Colombian encyclias or those of neighboring areas, and I have been in the habit of referring to it as the Unknown Encyclia. I now give it a proper Latin name. The flowers are usually turned up in the displays put on by the Colombian Orchid Society, and I have often wondered if the flowers were from different plants or the same one as I saw them on the various occasions.

The rather fleshy sepals and petals are red-brown over a greenish base color and have a somewhat glaucous appearance, even on the lip. The tips of the sepals and their bases may show more green. The lips have bright purple veins on a green-white background, sometimes with a rosy flush. The midlobe has a narrow whitish margin, and the sepals and petals may have a thin greenish edge.

In a show in Quito, the Colombians had a fine specimen of this plant, but they had no real name for it and called it "*Encyclia oncidioides* var. *expansa*" because of its large panicle. That specific name has never been published to my knowledge, and the species should not in any way be connected to *Enc. oncidioides. Encyclia expansa* (please see) is a completely different species. When I asked about its original location in Colombia, I was told somewhere near the river Magdalena. That is a long river through Colombia and does not tell us much, but it makes a good species epithet for that country until more specimens can be authenticated. I have not identified any to date in any of the major herbaria.

Although the lip features of *Encyclia magdalenae* are similar to those of *Enc. replicata* or *Enc. maderoi,* the pigmentation is completely different and the panicle habits are different—this species having broadly expanded side branches. The plants of most encyclias are similar, but vary much in size according to their habitats and exposure to sunlight. These plants are large with pyriform pseudobulbs and two or three leaves—not remarkable features in this genus.

The dorsal sepals measure 20 × 6.5 mm and are obovate with narrowed bases and acute tips. The petals are 19 × 7.5 mm, obovate-spatulate, with an apicule at each tip and a base somewhat clawed. The lip is 19 mm long and 21 mm across the flattened tips of the lateral lobes. The lateral lobes are veined at the base, greenish, 9 mm long × 4 mm wide, veined on the disc. The isthmus is 2 mm and clawed at the lip base about 3 mm. The column is 11 mm long, widened at the apex from a slender central region, and has auricles. The ovary is roughened, and with the pedicel measures 30 mm.

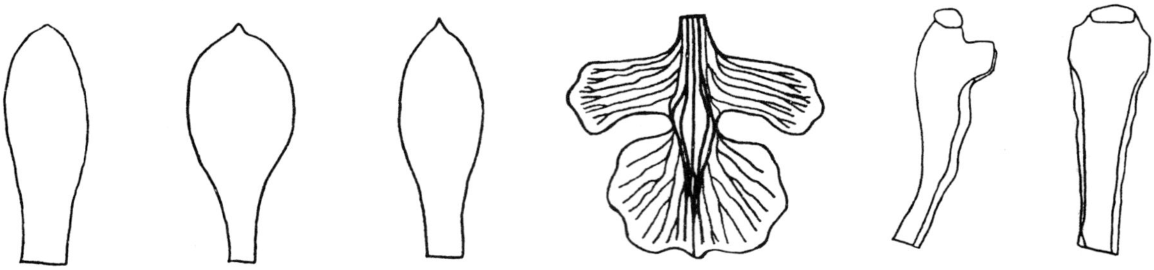

Figure 2-19. *Encyclia magdalenae.* This species has been displayed by Colombian orchid enthusiasts in shows from Florida to Guayaquil, but it has never had a proper name until now. Please see text for details. Drawing by Michael Ashe.

Giant-flowered Encyclia

Brazil

(FIGURES 1-4, 2-20; PLATE 42)

Encyclia megalantha (B. Rodrigues) Porto and Brade. 1935. *Rodriguésia* 1(2):29.

Synonyms

Epidendrum megalanthum B. Rodrigues. 1877. *Gen. & Spec. Orch. Nov.* 1:57.
Encyclia hollandiae Fowlie. 1990. *Orchid Digest* 54:157.

Misapplication (*auctor non*)

Encyclia megalantha var. *espiritusanctensis* Menezes. 1990. *Inst. Bras. Meio Amb. Comunicado Tech.*, p. 1 = *Enc. spiritusanctensis.*

Barbosa Rodrigues described this species as coming from Barbacena in Minas Gerais state, and it was flowering in January. He made a painting of it—in the field as Fowlie and Duveen (1992) have pointed out—and that painting was published in *Martius* by Cogniaux as Plate 20 (also see *Encyclia osmantha*). The painting and Rodrigues's type description tell us that the pseudobulbs were elongated and oval, carried two erect linear leaves, and the nodding panicle was well beyond leaf height. The sepals were lanceolate-oblong and acute. The petals were narrowly obovate, clawed, and obtusely pointed. The sepals and petals were brown with a narrow yellow margin and concave. The lip had heavy shoulders and lateral lobes with truncate to slightly widened and rounded scalloped tips; the midlobe had an indented tip, slightly undulate edges, and a distinct isthmus. The lip was pure white with carmine veining on both lateral lobes and midlobe. The slightly curved column was white. Measurements, as provided by Cogniaux in *Martius,* are pseudobulbs 8–10 cm tall; leaves 40–50 cm long × 2.5 cm wide; sepals to 24 × 9 mm; petals to 28 × 10 mm; lip to 30 mm long × 18 mm wide; and column to 13 mm long. The plants were epiphytic.

Figure 2-20. *Encyclia megalantha.* Type drawing by Barbosa Rodrigues published as *Epidendrum megalanthum* by Cogniaux. Plate 20 of *Martius* 3(5). This species has often been confused with *Encyclia osmantha* and other Brazilian species. Note the proportionately small apices of the lateral lobes and the pronounced notch at the apex of the lip.

Except for the plate of Rodrigues in *Martius,* the species has not otherwise been illustrated until recently by Sprunger et al. (1996), nor has it been clearly distinguished in cultivation. Fowlie and Duveen (1992) seem to have had difficulty with its identity, and they have proposed another name for the taxon, *Encyclia hollandiae,* based on plants from faraway Sergipe, about 1500 km east-northeast of Barbacena. Since we know little of the actual distribution of the Brazilian species, and there is still the proper identification of many herbarium specimens to complicate the matter, we do not know really whether the new name is necessary.

The color plate in Hoehne (1949) labeled *Encyclia megalantha* shows the orbicular petals more characteristic of *Enc. osmantha.* Fowlie and Duveen (1992) say *Enc. hollandiae* is distinguished by its obtuse petals; broadly ligulate lateral lobes of the lip with rounded apices; rounded, ovoid pseudobulbs; flowering in early Brazilian summer; and its Sergipe location. In contrast, *Enc. megalantha,* they say, has oblong, lanceolate, acute sepals; subquadrate lip lateral lobes with rounded apices; oblong, lanceolate, acute petals, and fusiform pseudobulbs. I have found, generally, that pseudobulb shape is not trustworthy as a species determinant. Shape may vary with the amount of light and environmental exposure of the plants, and with how crowded the pseudobulbs may be in a clump of plants. Obviously we need more field observations to resolve these details.

Menezes's Encyclia

Brazil

(FIGURE 1-4; PLATE 43)

Encyclia meneziana Gonzalez. 1992. *Bol. C.A.O.B.* 4:38. Reprinted in the *Orchid Digest* 57(1):45.

A confusing number of Brazilian encyclias have been published as a result of the work of Fowlie and Duveen in the past few years. They all need populations raised from selfed or sib-crossed seed to make them more readily available. This species honors one of the co-workers of Fowlie and Duveen, Lou Christian Menezes, a forestry engineer and ecologist, who has traveled extensively in Brazil and has written many articles and several books concerning Brazilian orchid species. She has helped me considerably in organizing the new material for Volume VI, has provided several of the color illustrations published here, and has generally been an informative and generous colleague. In Brasilia where she works when not on field trips, she has one of the most exciting orchid collections in that remarkable and vast country—the result of many hours of travel in out-of-the-way places as she carries out her job. The article cited above is written by Father Jose Gonzalez Raposo of Rio Claro City in São Paulo, who is a botanist dedicated to studying Brazilian orchids. He discovered the type specimen while collecting at 600 m in the municipality of Goiandira (Jaragua farm) in southern Goias state. The type specimen is conserved at the herbarium of the University of Brasilia, UB #27.

The species has short leaves and a flower panicle that reaches 1.5 m. The sepals and petals are brown on a green base that shows for the lower third of their length and around the segments as a narrow rim. The lip is white with green-brown flushes at the base and with purple veining about halfway down the lateral lobes. The callus is white marked with rosy veins that extend in papillate-verrucose fashion over the white disc of the midlobe. A wide white margin around the edges matches the terminal third of the lateral lobes. The column is white with rose flushes shading toward brown-green at its base, and has a maroon anther cap. Although the type description does not say, the column appears to have auricles, and the ovary surface appears smooth.

The sepals are 20 × 5 mm, the petals similar but 7 mm wide; both are oblong-elliptic; the petals more spatulate with a broad claw; the labellum with rounded midlobe 8 mm in diameter. No size is given for the lateral lobes, but the photos show them to be well shouldered, broad, and with rounded apices. The midvein of the lip forms a thickened ridge half the length of the midlobe, and the midlobe is concave around it. The species has not been in cultivation to date, except possibly in Brazil.

Pointed Tip Encyclia

Colombia, Ecuador, Peru

(FIGURES 1-4, 2-21; PLATE 44)

Encyclia microtos (Reichenbach f.) Hoehne. 1952. *Arq. Bot. Est. São Paulo* 2(6):152.

Synonym

Epidendrum microtos Reichenbach f. 1856. *Bonplandia* 4:326.

I am never certain that this species is different from *Encyclia aspera*, which also grows in Peru and Ecuador. They are certainly first cousins if not variants of the same species. I find, however, in making a list of the differences between them, that it is possible to make up a set of characteristics that do not seem to overlap. Time will tell if others can make further points in distinguishing the two species.

The most important difference that I have observed is the pointed tapered tips on the lateral lobes of *Encyclia microtos*. In fact, the name appears to have been derived from this quality, designating the pointed tips as a distinguishing feature. No other species of *Encyclia* has this character to the same degree. The lateral lobes of *Enc. aspera* are more obliquely acute or even rounded, and sometimes the edges roll under to make a more pointy and narrower appearance than the lobe structure shows when properly flattened. The flattened lip shows the lateral lobes of *Enc. microtos* pointing down about 45 degrees, and those of *Enc. aspera* extending more to the side. Also the flowers of *Enc. microtos* have six or seven major veins emerging from the forcipate callus ridges, while in *Enc. aspera* the callus tends to form three fleshy teeth that form three to five major veins. The anther cap of *Enc. microtos* is spotted with maroon or is dark in color, while that of *Enc. aspera* is yellow. On the column of *Enc. microtos* there are purple flecks, and the auricles tend to be small pointlike extensions, and those of *Enc. aspera* are larger and yellow in color. The distinctions for the two species are the sort that make one wish for a DNA test that could de-

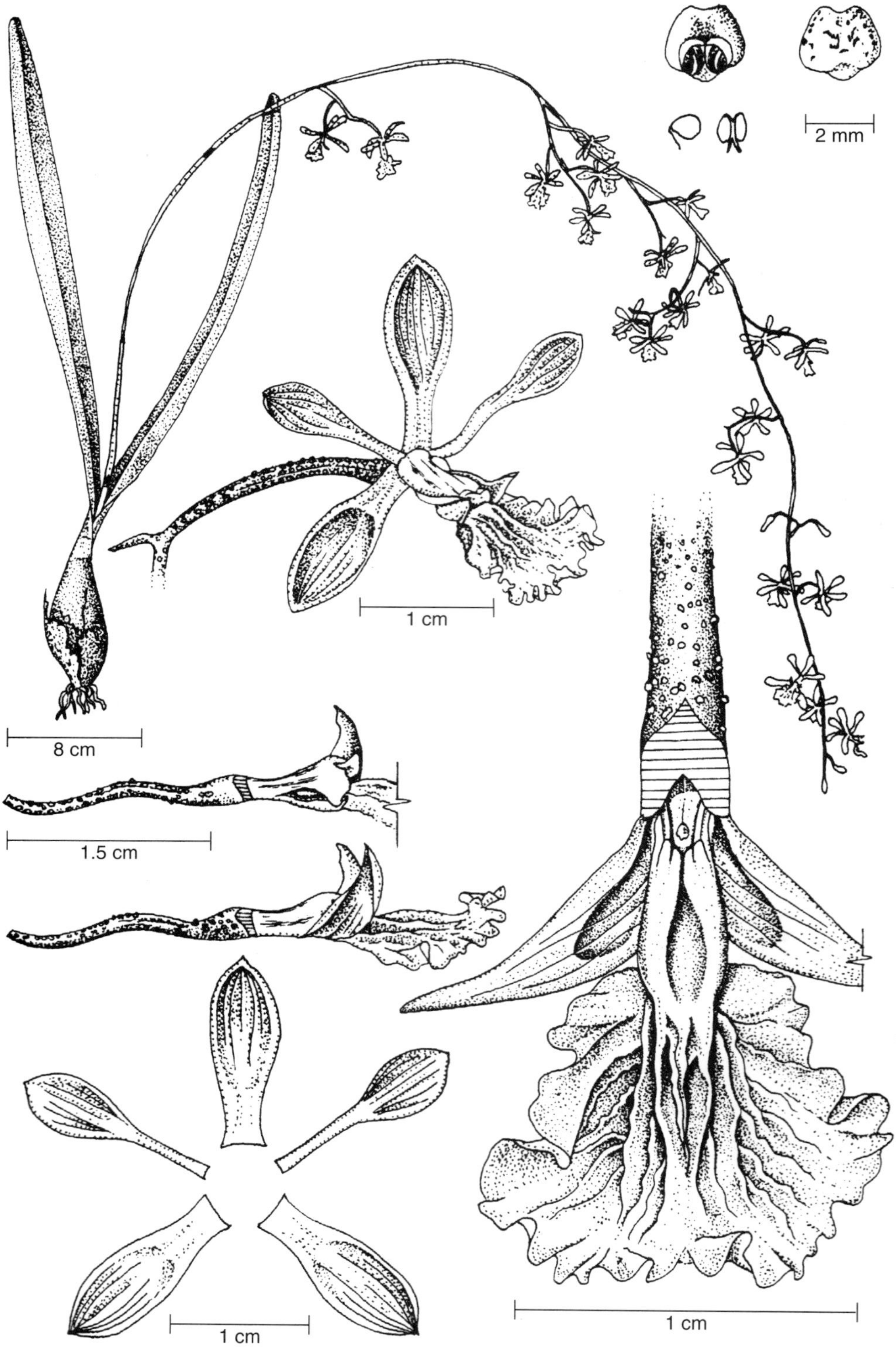

Figure 2-21. *Encyclia microtos.* Drawing by J. Perez C. published in *Icones Orchidacearum Peruviarum* by Bennett and Christenson. This species is sometimes mistaken for *Enc. aspera.* Note the pointed lateral lobes of the lip of this species, an unusual feature in this genus. With permission of the authors.

cide for once and all that the two were completely different—an end to such confusions! The petals are narrowly clawed.

Both species have a roughened, papillate ovary and flower stalk surfaces. Please see *Encyclia aspera* above. The type specimen for *Enc. microtos* was collected on the river Marañon by Josef von Warscewicz, #48, in Peru and is conserved in the Lindley Herbarium at Kew Gardens. A similar specimen appears in the Reichenbach Herbarium in Vienna. The flowers appeared in May. Measurements are as follows: dorsal sepal 14 × 4 mm; petals 13.5 × 4 mm; lip 13 mm with 1.5 mm clawed base, 2 mm isthmus, and wedge-shaped midlobe base 6 × 6 mm.

Marv Ragan of M & J Orchids, Florida, has received an AOS Cultural Award for his specimen plant of *Encyclia microtos.*

Small Yellow Encyclia

Brazil

(FIGURE 1-4; PLATE 45)

Encyclia microxanthina Fowlie. 1991. *Orchid Digest* 55(2):90. *Nomen nudum.*

As the species epithet implies, this plant has small yellow flowers. The plants that led to the description of *Encyclia microxanthina* were confiscated from a commercial shipment of orchids en route to Florida from Paraguay. The plants were then sent to the Los Angeles state and county arboretum for salvage where they sat under a bench waiting to be potted. Before that could happen, they began to send out flower stalks, and they flowered in October, 1990. Remaining material on their 2–3 mm wide roots suggested that they may have grown on rocks rather than bark. The pseudobulbs were ovoid-pyriform 3–4 cm tall, with one to two leaves 15–18 cm long × 7–8 mm wide. The leaves were succulent and stiff, shiny green, erect, and with a longitudinal groove, sharp apex and a twist, indicating to Fowlie a desert habitat. The flower scapes were 12–13 cm at leaf height and had three to five flowers with a natural spread of 2.8 cm.

The sepals were very narrow, acute, oblanceolate, to 14 × 3 mm; the petals were thinner and narrower than the sepals, 14 × 2 mm. Both were olive green. The flattened labellum measured 10 mm long × 9 mm across the lateral lobes, and the midlobe measured 5 mm long × 6 mm wide. The basal third of the lip was greenish, the balance an intense chrome yellow with a white callus. The lateral lobes were triangulate with an obtusely pointed apex at an oblique angle from the base. The midlobe was rounded to oval without a distinct apex but had distinct "corrugulose" veining. The anther was maroon or dark red-brown.

Unfortunately for us there is no Latin description of this species, no mention of a type specimen, or where the type specimen had been placed for conservation. That makes this species invalidly published, and we can only hope that the people at the Los Angeles Arboretum will remedy the situation for this distinctive and unique miniature encyclia. I have found in Fowlie's research notebook that the holotype specimen was #FR90P2 and that it was deposited in the UCLA Herbarium around 1990. Some botanist in Paraguay could rectify the situation and tell us more about the specimens and where exactly they grew. Also, we need to know if the lips were pubescent to any degree, the shape of the anther cap ("horned" or not?), and the ovary surface texture. We could then validate the name.

Naranjapata Encyclia

Ecuador

(FIGURE 1-5; PLATE 46)

Encyclia naranjapatensis Dodson. 1977. *Selbyana* 2(1):50.

This species without synonyms has a difficult name after a small Ecuadorian town where it was found. The town, named for a wide-spreading orange tree, is in the valley of the river Canar in Bolívar province and was just a crossroads that I have not yet found on a map. I once spent an afternoon there searching for this plant in vain. I had known of it since about 1965, when José Strobel sent orchid plants to the Brooklyn Botanic Garden. Some hobby growers in the New York area also ordered plants from Strobel, so this encyclia turned up in their collections. Strobel thought the species was new to Ecuador, and in that case suggested it be named for him. Somehow, it was never formally named. A few years after it was known as *Encyclia strobelii*, Cal Dodson in Ecuador was prevailed upon to give it a published name.

The plants are large and bold, and the large flowers in a tall panicle are white-lipped with red-brown to green-brown sepals and petals and a yellow border. The base of the midlobe has a few red-purple veins. Dodson measures the flower panicles as 70 cm tall. The pseudobulbs are clustered and 7 cm tall with two or three narrowly linear leaves to 60 cm long × 2 cm wide. The sepals are narrowly obovate, 20 × 6 mm, and the petals are spatulate-obovate, 20 × 10 mm. The lip is 20 mm long and has a 3 mm clawed base. The lateral lobes are falcate-oblique, veined about halfway, and the tips are white and rounded, somewhat flared. The midlobe is oval, slightly notched at the tip. The callus ends in three thickened veins. The column is 10 mm, greenish, has quadrate auricles, and a yellow-orange anther cap. The ovary surface is rough at the base but smoother toward the pedicel. The flowers have a distinctive fragrance.

Dodson comments that the plants grow in an area that is in the rain shadow of the mountains at elevations of 700–1000 m. The species appears to be endemic to the provinces of Bolívar and Canar in what are called the southern highlands of Ecuador. Plants do well on slabs or in baskets and should not be overwatered.

Oncidium-like Encyclia

Brazil, Guianas?, Guyana?, Suriname?, Venezuela?

(FIGURES 1-5, 2-22, 2-23, 2-24; PLATES 47, 48)

Encyclia oncidioides (Lindley) Schlechter. 1914. *Die Orchideen*, ed. 1, p. 210.

Synonyms

Epidendrum ensiforme Vellozo. *Icon.* 1827 (1831). *Fl. Flum.* 9, t. 2. Text 1881, L. Netto, ed. *Arch. Mus. Nac. Rio de Janeiro* 5:356. *Auctor non* Ruiz and Pavón = *Encyclia cyperifolia.*

Epidendrum oncidioides Lindley. 1833. *Bot. Reg.* 19, t. 1623.

Epidendrum longifolium B. Rodrigues. 1882. *Gen. & Spec. Orch. Nov.* 2:133. Also see Cogniaux in *Martius* 3(5):59, t. 18.

Epidendrum oncidioides sensu Cogniaux. 1898–1902. In *Martius* 3(5):47, p.pt. t. 11. *Auctor non* Lindley.

Encyclia longifolia (B. Rodrigues) Schlechter. 1914. *Die Orchideen*, ed. 1, p. 210.

Encyclia ensiformis (Vellozo) Hoehne. 1952. *Arq. Bot. Est. São Paulo*, n.s., 2(6): 151.

Encyclia vellozoana Pabst. 1975. *Bradea* 2(7):22.

Misapplications (*auctor non*)

Epidendrum affine Focke. 1853. *Bot. Zeit.* 11:341 = *Encyclia guianensis* Carnevali and Romero. (Please see text under *Encyclia guianensis* for *Epidendrum affine* Reichenbach f., 1856, and *Epidendrum affine* Richard, 1830, for other homonyms making name invalid.)

Epidendrum graniticum Lindley. 1841. (Please see *Encyclia granitica.*)

Epidendrum oncidioides var. *graniticum.* Two species combined together when there was no relationship except they both belonged in the same genus.

Encyclia granitica (Lindley) Schlechter in this case = *Encyclia oncidioides* (please see text). Identifier mixed up the two examples.

Encyclia oncidioides var. *granitica.* Reverse of above, but it was *Enc. granitica.*

Epidendrum profusum Rolfe. 1914. *Curtis's Bot. Mag.* 140, t. 8551 = *Encyclia profusa* (please see text).

Encyclia profusa (please see text).

Epidendrum perplexum = *Epidendrum oncidioides* var. *perplexum* Ames, Hubbard, and Schweinfurth. 1934. *Bot. Mus. Leaflt.*, Harv. Univ. 3(7):108 = *Encyclia perplexa* (Ames, Hubbard, and Schweinfurth) Withner. 1998. (Vol. V, p. 123, of that text.)

Encyclia perplexa (please see above).

Epidendrum oncidioides var. *itabirae* Zimmerman. 1934. *Biblio. Botanica* 109:6 = *Encyclia patens.*

Epidendrum oncidioides var. *gravidum* (Lindley) Ames, Hubbard, and Schweinfurth. 1935. *Bot. Mus. Leaflt., Harv. Univ.* 3(7):104 = *Encyclia gravida.*

Epidendrum oncidioides var. *ramonense* (Reichenbach f.) Ames, Hubbard, and Schweinfurth. 1935. *Bot. Mus. Leaflt., Harv. Univ.* 3(7):103 = *Encyclia ceratistes.*

Epidendrum oncidioides var. *mooreanum* (Rolfe) Ames, Hubbard, and Schweinfurth. 1935. *Bot. Mus. Leaflt., Harv. Univ.* 3(7):106 = *Encyclia mooreana.*

Epidendrum oncidioides var. *profusum* (Rolfe) Ames, Hubbard, and Schweinfurth. 1935. *Bot. Mus. Leaflt., Harv. Univ.* 3(7):106 = *Encyclia profusa.*

Epidendrum oncidioides var. *perplexum* (Rolfe) Ames, Hubbard, and Schweinfurth. 1935. *Bot. Mus. Leaflt., Harv. Univ.* 3(7):108 = *Encyclia perplexa.*

Encyclia oncidioides var. gravida (Lindley) Hoehne. 1952. *Arq. Bot. Est. São Paulo*, n.s., 2(6):153 = *Encyclia gravida.*

Encyclia oncidioides var. *ramonensis* (Reichenbach f.) Hoehne. 1952. *Arq. Bot. Est. São Paulo*, n.s., 2(6):153 = *Encyclia ceratistes.*

Encyclia oncidioides var. *mooreana* (Rolfe) Hoehne. 1952. *Arq. Bot. Est. São Paulo*, n.s., 2(6):154 = *Encyclia mooreana.*
Encyclia oncidioides var. *profusa* (Rolfe) Hoehne. 1952. *Arq. Bot. Est. São Paulo*, n.s., 2(6):154 = *Encyclia profusa.*
Encyclia oncidioides var. *perplexa* (Rolfe) Hoehne. 1952. *Arq. Bot. Est. São Paulo*, n.s., 2(6):154 = *Enc. perplexa.*

This famous and complicated set of names represents one of the early attempts to sort out the various encyclias of Central America, México, and South America. Ames and his coworkers, I can comment today with much hindsight, tackled this group of species as a portion of the genus *Epidendrum,* an enormously complex and time-consuming research project that resulted in their 1936 monograph, *The Genus* Epidendrum *in the United States and Middle America.* For additional useful and informative reading, *The Genus* Encyclia *in México* by Dressler and Pollard (1974) must also be consulted. That book, above all, gave *Encyclia* its present acceptance as a genus. It also demonstrated conclusively that several species from México and Central America were not varieties of the Brazilian *Enc. oncidioides,* but species in their own right. In view of this complicated history, I have included more figures and plates of this species than is usual. All orchid growers should learn to distinguish it quickly.

Additional specific history of the nature of the genus *Encyclia* is treated in Chapter 1 of Volume IV of this series, and the reader should see those pages for more detailed information. The reasons Ames and his colleagues transferred the epithets of several Central American and Mexican encyclias (epidendrums) to be varieties of a Brazilian species remain obscure except for the perplexity they complain about in their 1935 paper and the interpretation of overlapping characteristics. They felt that the "variability and close affinity" indicated "a single polymorphic species." We are still trying to distinguish all the species clearly from one another; but one thing we have definitely learned to date is that none of the Mesoamerican species are varietal forms of the Brazilian *Enc. oncidioides.*

I first became aware of the actual identity and qualities of *Encyclia oncidioides* when I studied an article by Robert Dressler in 1962, "Is it really *Epidendrum oncidioides?*" from the *Florida Orchidist.* We had been growing an unidentified Brazilian orchid at the Brooklyn Botanic Garden for a year or two. It had just flowered and was a perfect match for Dressler's drawings and remarks. That plant and his article really pushed me over the edge, and from that point on I became a devotee to the genus, gathering as much data on it as I could find. I was also working on the equitant oncidiums at the time, and what better excuse did I need to visit the Oakes Ames Orchid Herbarium at Harvard as often as I could get away from my professorial responsibilities at Brooklyn College. The Ames Herbarium has one of the best collections of *Encyclia* specimens in the world. In a large measure that article, the herbarium, and that plant of *Enc. oncidioides* in the Brooklyn Botanic Garden orchid collection have enabled me to write what is in this series of books today.

My notes at that time about the Ames record of the Lindley type specimen from Kew and a few other examples in the herbarium are as follows: no collector indicated on the type herbarium sheet; panicle of flowers strong, erect to 90 cm high, long-stalked; branches short and densely

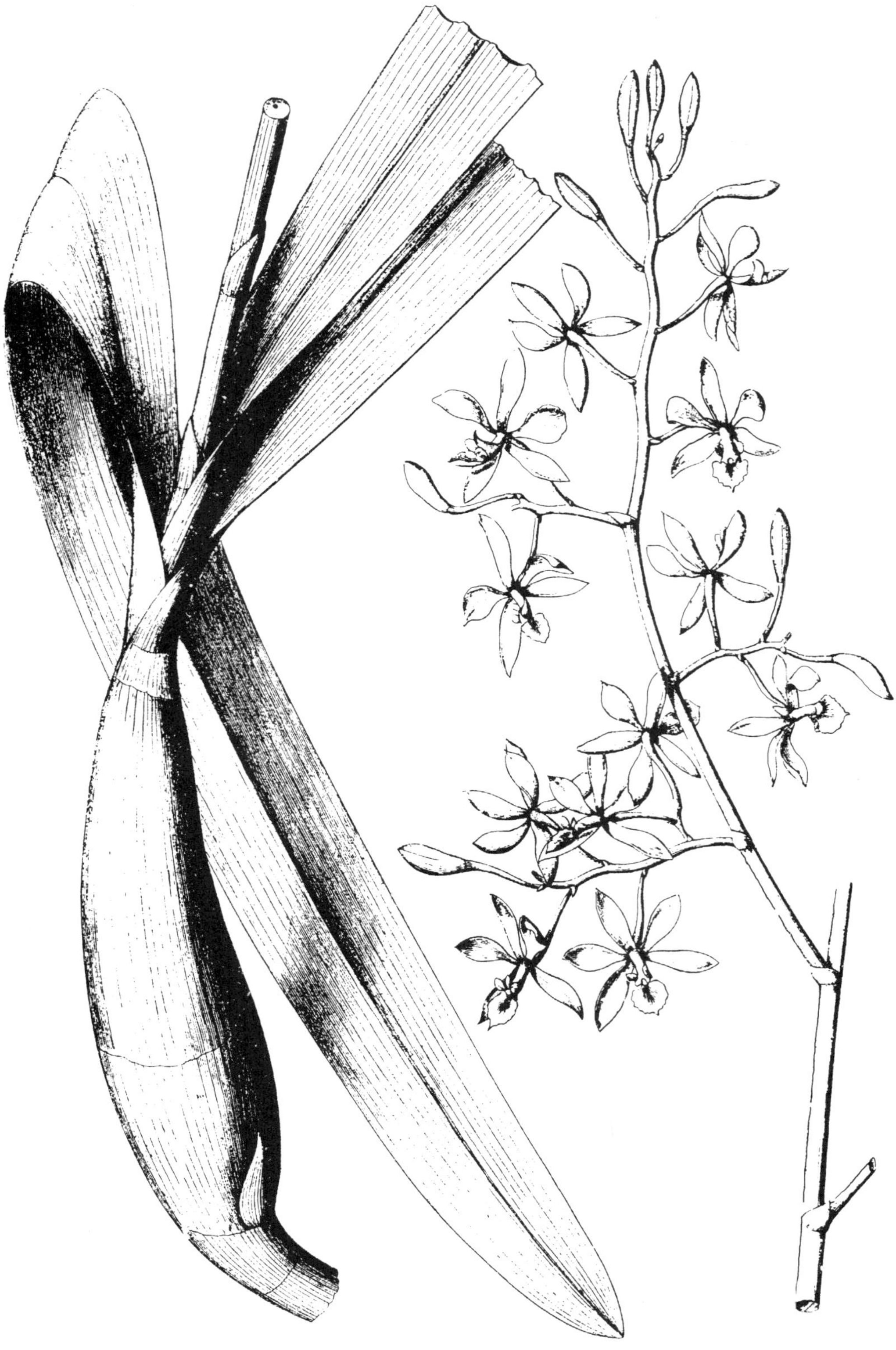

Figure 2-22. *Encyclia oncidioides.* Type drawing by Barbosa Rodrigues. Plate 11 of *Martius* 3(5), by Cogniaux. This species is often confused with others. Note the tall pseudobulb from an extended rhizome, the long leaves, and the erect flower stalk with densely placed flowers.

Figure 2-23. *Encyclia oncidioides.* Type drawing by Barbosa Rodrigues published as *Epidendrum longifolium* by Cogniaux. Plate 18 of *Martius* 3(5). See text for more details. Note the growth habit and flowers are the same as in Figure 2-12. Some say it is not the same species, but, if so, I have never been able to find another match for it among the many herbarium specimens I have examined. Perhaps another example may be found next week! It could happen.

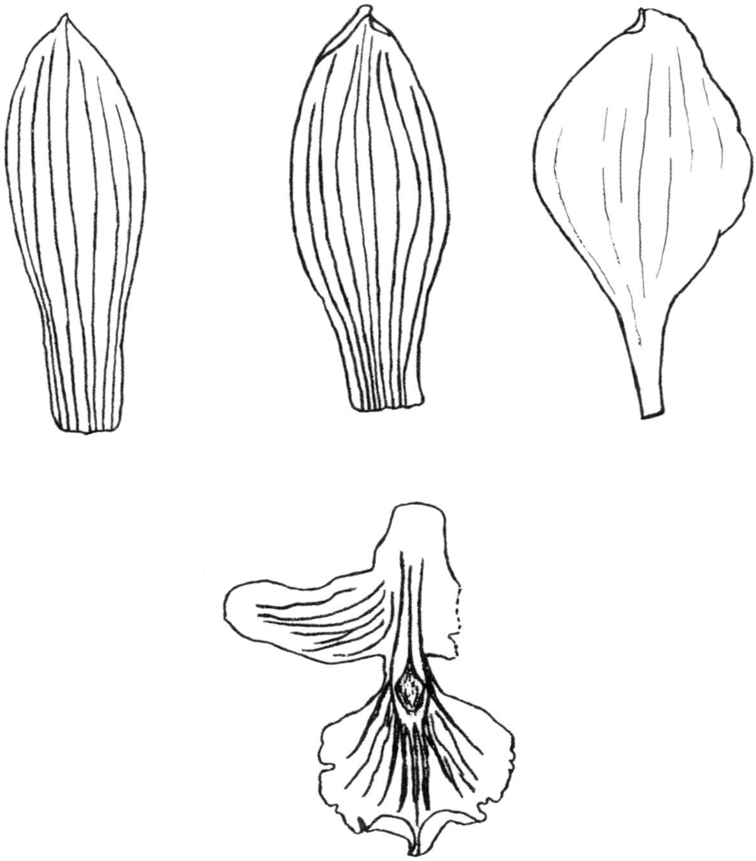

Figure 2-24. *Encyclia oncidioides.* A flower from the type specimen of *Encyclia oncidioides,* São Paulo, #2559, conserved in the Lindley Herbarium at Kew Gardens. *Top row, left to right:* dorsal sepal, lateral sepal, and petal. *Bottom row:* lip. Camera lucida drawing by Antonio Toscano do Brito.

spaced and flowered; sketch of lip and flower on type herbarium sheet comparable to view in the 1833 *Botanical Register* 19, Plate 1623 type drawing; spatulate rounded petals on properly identified specimens; lip acute; pseudobulbs coarse and heavy, 9 cm tall, separated by rhizomatous growths more than usual; leaves as many as four, 37 cm long × 3 cm wide; apparently rhizomatous and can grow on the ground. To this I would now add the following. The flowers are sweetly fragrant, like roses and violets, "delicious and powerful" it says in the *Botanical Register.* The petals are definitely clawed (have a narrowed stalklike base). The sepals and petals are a reddish orangey mottled brown, sometimes almost speckled, and with thin, distinct yellow edging. The lip is yellow to creamy white with an almost circular midlobe and a pointed tip. Red-purple markings are above the main veins of the lip, particularly the three central, somewhat raised veins. The column has distinct auricles, and the flowers are compact, middle-sized for the genus, and round in outline. My notes also show the column to be green with reddish veins toward its base and having a yellow anther cap.

Measurements show the sepals about 18 × 7 mm, more oblanceolate in shape than the spatulate petals that are about the same length but 9 mm wide across the more orbicular apex and only 2 mm wide at their base. The lip is 14 mm long and 7.5 mm across the midlobe. The lateral lobes are widespread from a broad base and have rounded apices without dark veining. They are 4.5 mm long. The isthmus with its forcipate callus is distinct between the base of the lip and the midlobe. The column is 6–8 mm long.

These large, strong plants apparently prefer a terrestrial or ruderal environment in the white sand areas that occur in various locations in northern Brazil. Such areas are called *caatingas,* and they also occur in southern portions of Venezuela and the Guianas. The white sands are probably the remains of old lake beds, and they usually have a unique flora compared to surrounding areas. We have similar regions in the United States in coastal New Jersey, the Pine Barrens, with a unique flora. They are also present in reduced amounts in New York State south of Albany. Romero (1993a, 1993b) has written about them in Venezuela where many such areas remain to be explored for orchids and other plants limited to such habitats.

In cultivation, at least, *Encyclia oncidioides* may flower in May to August, the peak being, according to Hamilton's data (1986), in June and July. Actual distribution depends in part upon proper identification of the species, and the species has often been confused with *Enc. granitica* (please see). The latter has a generally similar habit of growth and a preference for white sand areas or rocky outcrops. Plants of *Enc. granitica* are not as large, however, and the flowers are in nodding, open panicles with the sepals and petals showing distinct speckles or mottlings of dull, dark red-brown on a yellow-green background. These markings are not present in the orangey-brown *Enc. oncidioides* flowers.

A final point about one of the names is that *Epidendrum ensiforme* Vellozo from Brazil, transferred to *Encyclia* by Hoehne in 1952, is a homonymic name of *Epi. ensiforme* Ruiz and Pavón from Peru. Pabst noted this and published *Enc. vellozoana* in 1975 as a new name for the Brazilian species. Our modern synonymy above thus includes these epithets. Another problem area involves the epithet *affine* or *affinis* in this genus. Focke (1853), Richard (1850), Reichenbach (1856), and Schlechter (1918) have all used that epithet; Focke used it for this species.

The plants may be grown in large, flat, bulb pans, or they may be rooted on a horizontally suspended section of cork where they may clamber about. The type specimen is represented by Plate 1623 in the *Botanical Register*, 1833. The plants of this species, along with those of *Encyclia pachyantha,* are among the largest in the genus. Castro (1998) considers *Enc. longifolia* a synonym of *Enc. oncidioides.*

Capart's Scented Encyclia

Brazil

(FIGURES 1-5, 2-25; PLATE 49)

Encyclia osmantha (B. Rodrigues) Schlechter. 1914. *Die Orchideen*, ed. 1, p. 210.

Synonyms

Epidendrum osmanthum B. Rodrigues. 1882. *Gen. & Spec. Orch. Nov.* 2:134.
Epidendrum capartianum L. Linden. 1891. *Lindenia* 7:93, t. 333.
Epidendrum godseffianum Rolfe. 1892. *Gard. Chron.*, n.s., 1:136.

Now at last I must make up my mind in a controversy that has gone on for a long time—the sameness or differentness between *Encyclia osmantha* and *Enc. megalantha.* The two have been confused for years because of their similarities, which are complicated today because of three new taxa described by Lou Menezes, *Enc. lutzenbergeri* and *Enc. lutzenbergeri* var. *major* (please see), and *Enc. spiritusanctensis. Encyclia hollandiae* by Fowlie (1990) also enters the picture. We must go back and review carefully the type descriptions and the records, if any, that may still exist of the original plants and flowers. Dr. Capart was a well-known physician and orchidist of the time who had flowered his plant two years after receiving it.

Plants of *Encyclia osmantha* were found and named by J. Barbosa Rodrigues, the name meaning fragrant or scented. They were growing on the trees of forests in Minas Gerais near the Casa de Pedra to St. José d'El-Rey. They were flowering in September, were very aromatic, had brown sepals and petals washed with yellow-green, and had yellow-green lateral lobes veined in carmine and a white midlobe rayed in carmine at the base and washed with rose. The plants had two leaves, the sepals were oblong, subacute, narrowed at their base and, with the petals, curved in so the flowers appeared cupped. The petals were widely ovate-spatulate, somewhat acute but broader than the sepals at their tips, and their bases were narrowly and shortly clawed. The lip lateral lobes were oblong and then spreading and subrotund with crenate edges at the tips. The midlobe was also subrotund with a coarsely undulate margin that curved inwardly, but the halves of the lip reflexed to a degree along the central rib. The column was auricled and the ovaries were smooth.

The flowers were clearly portrayed in table 7792 of *Curtis's Botanical Magazine* (1901). In writing the description to accompany the plate, J. D. Hooker listed *Epidendrum godseffianum* and *Epi. capartianum* as synonyms. He also stated that the plant pictured was imported from Pernambuco by Sander and Company in 1899, and that the specimen was purchased for the collection at Kew where it flowered in the Cattleya House in 1900, the flowers lasting two months. The fleshy flowers were about 5 cm across. Wow! Besides the size and usual cupping of the flowers, I particularly note the hallmark subrotund, crenate, white-tipped, and pink-flushed lateral lobes of the lip, and the generous carmine veining on a white midlobe. The *Lindenia* plate and description match that of Hooker's plant and Rodrigues's description.

At Kew is the type specimen for *Epidendrum godseffianum* named by Rolfe and an interesting 1893 note:

> When I described *E. godseffianum,* I overlooked *E. osmanthum.* Afterwards a plate was sent by M. Linden asking me to describe it as *E. capartianum,* which I declined on the ground that it was already described. Then it was sent to M. Cogniaux, who in ignorance prepared the description, as I found when the proof of that part of *Lindenia* was sent to me. Immediately I wrote M. Cogniaux the facts, and he then withdrew his name. But the description appeared signed L.L.

The next time your plant of *Encyclia megalantha* flowers, check it out. Mine have always turned out to be plants of *Enc. osmantha* that have been mistakenly labeled. The lateral lobes of the lip, large and scalloped or frilly, will almost tell you by themselves. Plate 19 in *Martius,* fig. II, does not match the other examples of *Enc. osmantha,* the lateral lobes of the lip being

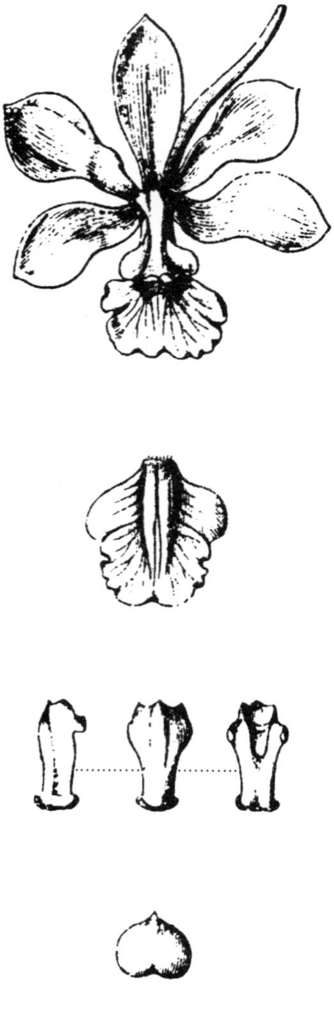

Figure 2-25. *Encyclia osmantha.* Type drawing by Barbosa Rodrigues published as *Epidendrum osmanthum* by Cogniaux. Adapted from Plate 19 of *Martius* 3(5).

too small. It is a better match for *Enc. megalantha* in Plate 20 of *Martius*, and I have wondered sometimes if it was mislabeled or just misplaced since they are on adjacent plates. Further, after reading over Fowlie and Duveen (1992) for the umpteenth time, I am unable to understand their treatment of this species and its synonyms. I am afraid also that they may have misled Lou Menezes to publish her *Enc. spiritusanctensis* that to me is a fine match for *Enc. osmantha.* Also, the type records and the note from Rolfe seem good evidence to me for the account accorded these species in this volume. Please consult Figures 1-4 (*Enc. megalantha*), 1-5 (*Enc. osmantha*), and 1-6 (*Enc. spiritusanctensis*) to compare the flattened lips of some of these large-flowered species that seemed to have been confused with one another.

There is some variation in the size of flowers from various clones of this species. One that I have flowered measured as follows: dorsal sepal 23 × 9 mm; petals 23 × 13 mm the claw 5 × 2 mm wide; the lip 23 mm long with a basal claw of 3 mm; isthmus 2 mm long, and midlobe 16 × 12 mm, distinctly notched. The lateral lobes are 12 mm long × 5 mm at their bases and 9 mm wide at their apices. The column is 10.5 mm long, and the pedicellate ovary 30 mm.

Fleshy-flowered Encyclia

Brazil, French Guiana, Guyana, Venezuela

(FIGURES 1-5, 2-26; PLATE 50)

Encyclia pachyantha (Lindley) Hoehne. 1952. *Arq. Bot. Est. São Paulo* 2(6): 154.

Synonyms

Epidendrum pachyanthum Lindley. 1838. *Bot. Reg.* 24:31, misc. 42.
Encyclia alboxanthina Fowlie. 1990. *Orchid Digest* 54(1):26.

Misapplications (*auctor non*)

Epidendrum pachyanthum Schlechter. 1919. *Repert. Spec. Nov. Regni Veg.*, Beihefte 6:38.
Encyclia hartwegii Lindley in Foldats's Orchidaceae in Lasser's *Flora de Venezuela* 5(3):268.

One of the major characteristics of this species is its large size. The type specimen has two or three leaves to 51 cm long with a pseudobulb about 10 cm tall, one of the largest plants in the genus. Lindley, describing the type flower of *Epidendrum pachyanthum,* remarked on the large size of the flowers being a full 5 cm in diameter, thick and fleshy, and dull green stained with dirty reddish brown, especially towards the ends of the sepals and petals. The lip was a pale straw color streaked in the middle with violet. The petals were not clawed. Lindley's plant was found wild in the interior of Demerara (now Guyana) by Schomburgk. It had no collection number and apparently had been growing at Loddiges's as their name is also on the herbarium label. As with many of these species with few herbarium specimens for the record, there is a question over identity and synonymy with other species. At Kew are a useful short note and a picture of a flowering branch (*Orchid Review,* p. 261, 1970) that shows many of its qualities.

Carnevali and Ramírez (1988) point out that *Encyclia pachyantha* may not be the same species as *Enc. remotiflora* or *Enc. latipetala* (please see), and I am considering that all three are different—in spite of their somewhat similar flowers and vegetative habit. Schlechter transferred the species from *Epidendrum* and then named a second plant with the *Epi. pachyanthum* epithet. This was a legal procedure, nevertheless somewhat confusing, and he thus avoided what would have been another homonymic name in the genus *Epidendrum.*

In the type description Lindley describes the leaves as straplike and wavy-edged, and their apices as oblique and rounded. The leaves are thinner and broader than usual for encyclias. A point I note in the greenhouse is that the leaves show an abscission zone, but it does not function. The dead leaves do not fall from the plant but must be cut away. The petals, not being clawed except at their very base where they have a narrow portion to attach to the receptacle, are obovate-lanceolate, somewhat spatulate, while the sepals are somewhat narrower and more pointed. The sepals

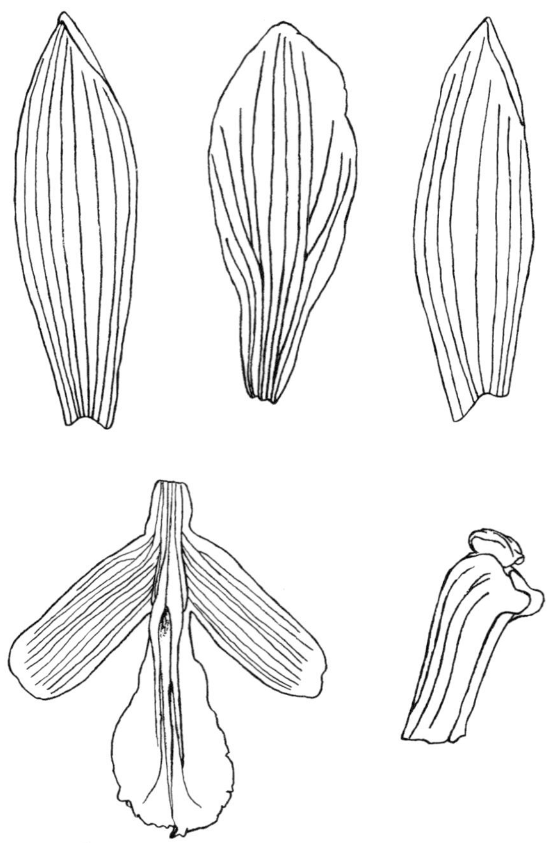

Figure 2-26. *Encyclia pachyantha.* A flower from the type specimen of *Epidendrum pachyanthum* sent by Richard Schomburgk to Loddiges and conserved in the Lindley Herbarium at Kew Gardens. *Top row, left to right:* dorsal sepal, petal, and lateral sepal. *Bottom row:* lip and column. Camera lucida drawing by Antonio Toscano do Brito.

and petals are basically an olive green and then have flushes of brown over the veins and especially toward their apices. The lateral lobes of the lip have a ligulate-oblong shape with white rounded ends that are clear of colored veining. The white to cream midlobe is broadly rounded-ovate. The central main vein, and sometimes the two laterals, are red-purple for a short distance just below the callus. The flower parts measure about 22–23 mm long, and the sepals and petals are both seven-veined. The lip is 20 mm long. The column is about 11–13 mm long, bright green at the base and golden yellow at the apex including the auricles and the anther cap. The ovary surface is smooth. The flower stalk is stiffly upright above the leaves and is sometimes not branched. The flowers are strongly scented, and the plants are summer bloomers in cultivation in the United States. They prefer a basket with osmunda, or a cork slab, keeping in mind their potential size, larger even than a well-grown plant of *Encyclia oncidioides.*

Plants of *Encyclia pachyantha* appear stately and will serve well as parents for making hybrids. At one point Orquidário Binot was marketing a hybrid of this species with *Cattleya forbesii,* but I do not believe the hybrid was ever named or registered. The flowers took after the *pachyantha* parent in coloration and shape, but had more veining marked on the lip. The massive lateral lobes were each larger than the midlobe—perhaps not surprising considering the dominance of the lip characteristics of *Cattleya forbesii* in hybrids.

Spreading-flowered or Sweet-scented Encyclia

Brazil

(FIGURES 1-5, 2-27; PLATE 51)

Encyclia patens Hooker. 1830. *Curtis's Bot. Mag.,* t. 3013.

Synonyms

Epidendrum odoratissimum Lindley. 1831. *Bot. Reg.* 17, t. 1415.
Exophya patens (Hooker) Rafinesque. 1836 (1838). *Fl. Tellur.* 2:63.
Epidendrum glutinosum Scheidweiler. 1843. *Allgem. Gartenzeit.,* p. 110.
Epidendrum serronianum B. Rodrigues. 1877. *Gen. & Spec. Orch. Nov.* 1:50.
Encyclia odoratissima (Lindley) Schlechter. 1914. *Die Orchideen,* ed. 1, p. 210.
Encyclia serroniana (B. Rodrigues) Hoehne. 1952. *Arq. Bot. Est. São Paulo* 2(6):155.

Misapplications (*auctor non*)

Epidendrum patens Swartz. 1797. *Flor. Ind. Occ.* 1:1495. Illustrated in *Bot. Cabinet* 16, t. 1537. Jamaica, Trinidad, West Indies = *Epidendrum fucatum* or *Epidendrum anceps.*
Macradenia lutescens Loddiges. 1830. *Bot. Cabinet,* 16, t. 1556. Trinidad = ?*Encyclia maravalensis* Withner (see Vol. IV).

This species had a tough beginning being confused with other species and having a name change, but we can now be correct in referring to it as *Encyclia patens* Hooker. Lindley, being very exact with Hooker's definition of the genus *Encyclia,* believed that this species belonged best in *Epidendrum.* Lindley (1831) also pointed out that the plate made from the Loddiges's plant of *patens* and published under the epithet *Macradenia lutescens* was a mistake in labeling as the *Macradenia* was a species of the Vandeae!

Since Hooker's *Enc. patens* could not be *Epi. patens* because of Swartz's previous use of the *patens* epithet for a West Indian *Epidendrum* in 1797, Lindley took the new name of *Epi. odoratissimum* to resolve the difficulty. But, the name *patens* was not in prior use in the genus *Encyclia,* so that combination becomes legal, and the *odoratissimum* basionym becomes unnecessary and therefore a synonym. Schweidweiler was also aware of these difficulties and came up with yet another name, *Epi. glutinosum,* in 1843.

The flowers were heavily scented, though other species in this genus also have scented flowers. It would seem a poor choice to ever name a species after its fragrance, especially the "most fragrant," as several such species epithets have ended up as synonyms or otherwise being changed. The Harrison plant pictured and described in Plate 3013 of *Curtis's Botanical Magazine* in 1830 as *Encyclia patens* is the type specimen. The plant in Plate 12 of *Martius,* published as *Epidendrum serronianum* in 1898, is notable for its three anthers, of which only the central one is functional, and its enlarged ovaries. The plate pictures an unusual and unique form, although it seems to have formed a population in nature. It was originally called *Epi. serronianum* (please see) by Barbosa Rodrigues. Most, today, consider

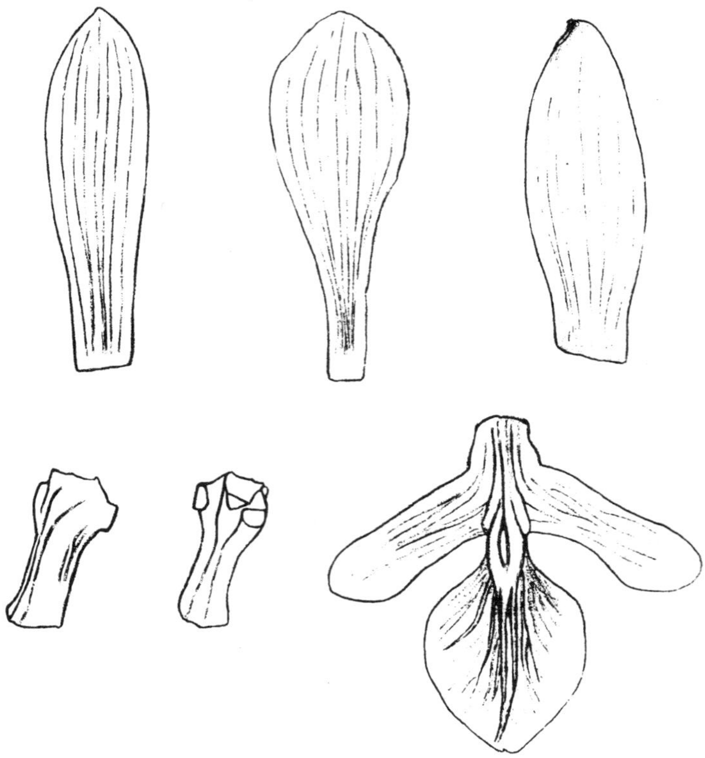

Figure 2-27. *Encyclia patens.* A flower from the type specimen of *Epidendrum odoratissimum,* a synonym, conserved in the Lindley Herbarium at Kew Gardens. *Top row, left to right:* dorsal sepal, petal, and lateral sepal. *Bottom row, left to right:* column from above, column from below, and lip. Camera lucida drawing by Antonio Toscano do Brito.

it an unusual form of *Enc. patens,* though to my knowledge no one has it in cultivation so we could study it further. It is equivalent, I believe, to the *Epi. cochleatum* var. *triandrum* population in Florida. The three anthers reflect an ancestral condition in orchid evolution when the basic flower had six stamens (Withner, 1974, chapter 6).

Another point to make, going back to the discussion of *Encyclia* as a genus in Volume IV, p. 13, is that Rafinesque in 1836 (1837) proposed this species, *Epidendrum patens,* as the type for his genus *Exophya.* I am pleased at this point that transferring all the species of *Encyclia* in these volumes to *Exophya* is not necessary.

Meanwhile, to describe the plants and flowers of *Encyclia patens,* it is easiest to return to the type description by Lindley in *Curtis's Botanical Magazine* and paraphrase his remarks. The flowers were described as dingy yellow-green (olive) within with purplish brown streaks, the three sepals lanceolate, and the petals spatulate and acutely pointed, the sepals and petals having a thick, fleshy texture and yellowish margins. The lip stands forward longer than the perianth segments, is narrow at the base, pale yellow streaked with red lines within, especially with a short line near the apex. The lateral lobes curve upward and include the column, and the midlobe is obovate and entire with the edges recurved. The column is semi-cylindrical, white, with two incurving auricles. The plant comes from the woods near Rio de Janeiro, was gathered by a Mr. Warr, and is very fragrant.

This, the second species of *Encyclia* ever described, was sent to Mrs. Arnold Harrison, sister-in-law to the collector William Harrison, who found the type specimen of *Enc. viridiflora* that has never been found again. That type, so far as I can tell, was undated and became the type for the genus *Encyclia* described by Hooker in 1838. Mrs. Harrison sent the specimen to Lindley with a sketch made in her

> rich collection at Aigburgh. The general structure of the flowers is so similar to those of the *Encyclia viridiflora,* communicated by the same lady, and from the same country, that I have no hesitation in referring it to the genus: the character of which, I find it consequently necessary in some degree to correct.

The upright somewhat stiff panicle has several closely flowered branches and is only a little taller than the two or three leaves. The leaves are a dull green and tinted with maroon flushes in the light. The plants flower in cultivation from late winter into the summer and lend themselves to basket or slab culture. The flower measurements are as follows: dorsal sepal 13 × 4 mm; petals 13 × 7 mm; lip 13 mm long, 17 mm spread between tips of lateral lobes; lateral lobes 8 × 2 mm widening toward 37 mm wide, rounded tips; midlobe 6.5 mm wide and long with slight broad point; column 10 mm; ovary and pedicel 16 mm.

Few-flowered Encyclia

Brazil

(FIGURES 1-5, 1-28; PLATE 52)

Encyclia pauciflora (B. Rodrigues) Porto and Brade. 1935. *Rodriguésia* 1(2): 29.

Synonym

Epidendrum pauciflorum B. Rodrigues. 1877. *Gen. & Spec. Orch. Nov.* 1:49.

Even though this species has comparatively large nodding flowers coming from a relatively small two-leaved plant, it has apparently been seldom cultivated. Rodrigues collected it in Minas Gerais state flowering in December. His plate of the species is in t. 16, fig. 1 of Cogniaux's account in *Martius*. It is also notable that the sepals and petals reflex, curving backward from the column and lip. The lip also curls down a little along the edges and at the tip. Only four or five flowers are produced on a short raceme. Rodrigues states that the sepals and petals are green-brown, the median lobe of the lip is white lined with purple, as are the lateral lobes that emerge from the green base of the lip. Some clones have a pink flush on the lip, and the lateral lobes can be yellowish; other clones have been described as having slate-blue or cerulean-blue lips, a unique color for this genus.

Measurements and shapes include pseudobulbs pyriform, about 2–3 cm long; leaves lanceolate, 10–13 cm long × 6–8 mm wide; sepals 22 × 6 mm, lanceolate and pointed; petals a little shorter, subspatulate, and distinctly clawed; lip 25 mm long with midlobe 12–13 mm wide, longer than broad, with a flattened apex and ruffled edges. The obovate shape—broader across the apex than at the base—of the midlobe is characteristic and unusual in the whole genus. The callus has three teeth at its tip, the central continuing as the central rib of the midlobe. The veining on the disc is granular-verrucose as though beaded. The whitish column is about 8 mm long with a white anther cap. The ovary surface is smooth.

The plants are epiphytic and grow well in small baskets with osmunda fiber. So far as I can tell, this species has never been judged or hybridized.

Pedra Azul Encyclia

Brazil

(FIGURE 1-5; PLATE 53)

Encyclia pedra-azulensis Menezes. 1992. *Orchid Digest* 56(3):148.

This relatively recently described species has no known synonyms. The type plant was collected at about 600 m elevation on Pedra Azul in Espírito Santo state of Brazil. It was found by Bernardino Amiao, a member of the Brasilia Orchid Society, and it flowered in Menezes's collection in Brasilia in November, 1991. The type specimen is conserved at the herbarium of the University of Brasilia.

The description says it is similar to *Encyclia osmantha* and is close to

Figure 2-28. *Encyclia pauciflora.* Type drawing by Barbosa Rodrigues published as *Epidendrum pauciflorum* by Cogniaux. Plate 16 of *Martius* 3(5). Conserved at Kew Gardens.

Enc. lutzenbergeri. The pseudobulbs are ovate-elongate, 9 cm tall × 4 cm wide, and with one to three acute leathery leaves, 50 cm long × 2.5 cm wide. The branched flower scape is 107 cm long with 44 fragrant flowers. The sepals and petals are green flushed with brown; the lip creamy white with verrucose veining in purple; and the lateral lobes have greenish bases and are darkly veined with white apical areas.

The sepals are 22 × 7 mm, the petals 20 × 7 mm, and the column 10 mm. This species differs from related forms by having the sepals and petals alike in lanceolate shape, the petals not spatulate. The midlobe of the lip is rounded with a ruffled edge, and the lateral lobes are broad and long, almost triangular, not strap-shaped as in the related species. The lateral lobes broadly extend, when the lip is flattened, to the sides of the midlobe.

Painted Encyclia

Brazil?, Guyana

(FIGURES 1-5, 2-29)

Encyclia picta (Lindley) Hoehne. 1952. *Arq. Bot. Est. São Paulo* 2(1):154.

Synonym

Epidendrum pictum Lindley. 1838. *Bot. Reg.* 24:32, Misc. no. 43.

Encyclia picta was named by Lindley in 1838 from a plant at Loddiges's greenhouses in England that came from Demerara (now Guyana). Lindley said it resembled *Epidendrum odoratum (patens)* and was closely allied to *Epi. chloranthum.* His drawing of the type specimen was a poor one, comparatively, and looking at it today I cannot see what he must have seen.

Encyclia picta has been a poorly known species. Lindley calls it "another of those unpublished species with dull yellow flowers neatly striped with crimson." The two flowers of the type are conserved at Kew. Lindley's sketch of a partly flattened lip and another of a whole flower are not very detailed nor satisfactory, though the petals are labeled dull ochre-yellow, and the sepal is labeled brick red over the veins at the apex. A pocket on the type sheet has only the two small flowers. The type description does not mention column appendages. In other words, there is little to work from for a large group of complicated species. Lindley does say that the leaves are leathery, bluntly rounded at their tips, and that the plant produces an upright panicle. The sepals and petals are nearly similar in shape, obovate-linear; the lateral lobes linear, acute, and subfalcate; the midlobe plicate-edged, ovate, acute, with raised veins. The column configuration and surface of the ovary are not mentioned.

Cogniaux also describes the species in *Martius* and provides some sizes and comments. The flowers are small. The sepals are five-nerved, little narrowed at their base, 16 × 3 mm. The petals are somewhat oblique, five-nerved, attenuate below, 16 × 3.5 mm. The lip is many nerved, 14 mm long and 5 mm wide across the midlobe. The column is slightly curved, 8–9 mm long. The petals may be flushed with red-brown over the veins, and the color may be concentrated toward their apices.

The previously unpublished Dunsterville drawing #852 made January, 1964, was considered by Carnevali and Ramírez (1988) to be *Encyclia*

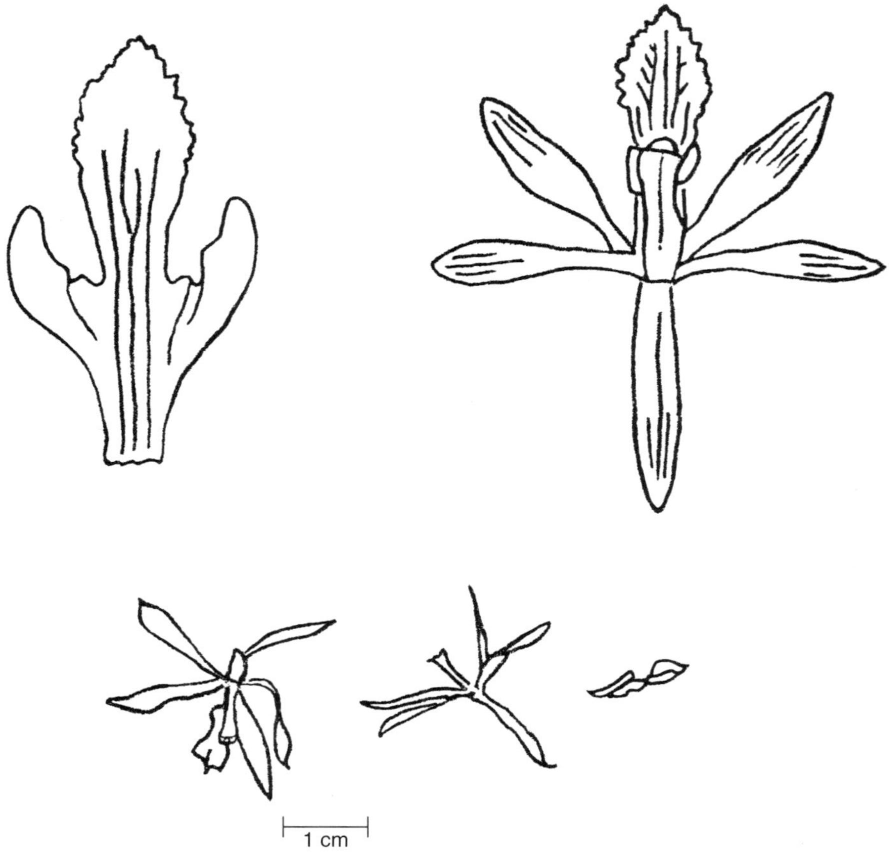

Figure 2-29. *Encyclia picta.* Drawing of the type specimen traced by the author at the Lindley Herbarium in Kew Gardens. The two flowers are the type specimen, and the type description is all that is known about the species.

picta, but six years later Carnevali et al. (1994) changed their minds and preferred to identify the drawing as *Enc. guianensis,* a previously unnamed species (please see). They comment that the small plants are less than 40 cm high with densely flowered panicles, and that the flowers are more membranous than those of other encyclias, a combination of traits unlike other Venezuelan encyclias. Their collections were from Bolívar state at 500–800 m and were flowering in December and January. The plants were intermingled in their habitat with those of *Enc. chloroleuca,* but easily distinguished. The flowers, according to Carnevali et al. (1994), are pictured under the incorrect name of *Enc. oncidioides* in the book by Werkhoven (1986) on Suriname orchids. I feel that Carnevali and his associates were correct in renaming the Dunsterville drawing even though there is still no new match for the Lindley *Enc. picta* concept, and the species remains somewhat of a mystery. Please see Figures 2-37 and 2-29 regarding *Enc. picta.*

Hairy Encyclia

Colombia, Peru

(FIGURES 1-5, 2-30; PLATE 54)

Encyclia pilosa (Schweinfurth) Carnevali and Ramírez. 1988. *XLVI Exposicíon de Orquídeas, Comité de Orquideología,* Bol. No. 23:13–87.

Synonym

Epidendrum pilosum Schweinfurth. 1943. *Amer. Orchid Soc. Bull.* 12:361.

I thought originally that this was *Encyclia leucantha* because of the hairy pubescent callus structure, but as time went on I could not place a detailed drawing I had made of a Kew specimen from the Bentham Herbarium. I finally realized the specimen was *Enc. pilosa* with its long and droopy lateral lip lobes that come to a blunt tip and extend to a position about to the midlobe on a flattened lip. This shows clearly in Figure 2-30, as do the short hairs that are on the lateral lobes and the callus and isthmus of the lip. According to Schweinfurth, the specific name alludes to the mat of hairs on the lip. The type specimen was collected by G. Klug, #779, at Loreto Mishuyacu, near Iquitos at 100 m elevation in January, 1930. It is conserved at the U.S. National Herbarium as specimen #1455771.

The type description says the plants are rather small, epiphytic, about 23 cm high with stout roots. The pseudobulbs are narrowly pear-shaped and have a single, narrow, linear leaf. The flowers, in a raceme or a sparsely branched panicle, are in groups of 8–10 and are small and brownish yellow-green. The dorsal sepal is oblanceolate-oblong, 13.2 mm long × 4 mm wide. The petals are 12 × 4 mm. The lateral lobes of the lip extend beyond the tip of the column and are obliquely ovate narrowing to the small blunt apex. The flat midlobe is 5 mm long × 6 mm wide and oval-orbicular. Five verrucose veins extend from the hairy callus ridges, the central three forming a small apicule at the apex of the lip. The column is arcuate, about 7.5 mm long, and has slight wings with some widening opposite the stigma—but not enough to be called auricles. In the plate with the type description, the anther cap appears to be "black" and deeply divided, but no mention of its color or form is in the description. The ovary is verruculose.

Although it has been collected once or twice, this species is not well known. A few examples should be brought to cultivation for seed production and propagation.

Piracanjuba Encyclia

Brazil

(FIGURE 1-5)

Encyclia piracanjubensis Menezes. 1991. *Bol. C.A.O.B.* 3(3):16.

This species is another of the new taxa coming from Brazil that are, as yet, poorly known beyond their type descriptions and type specimens. Lou Menezes has served as a clearinghouse for many of these species as a result of her many journeys around Brazil as a forestry engineer and her general knowledge of the various orchids found in the widespread regions.

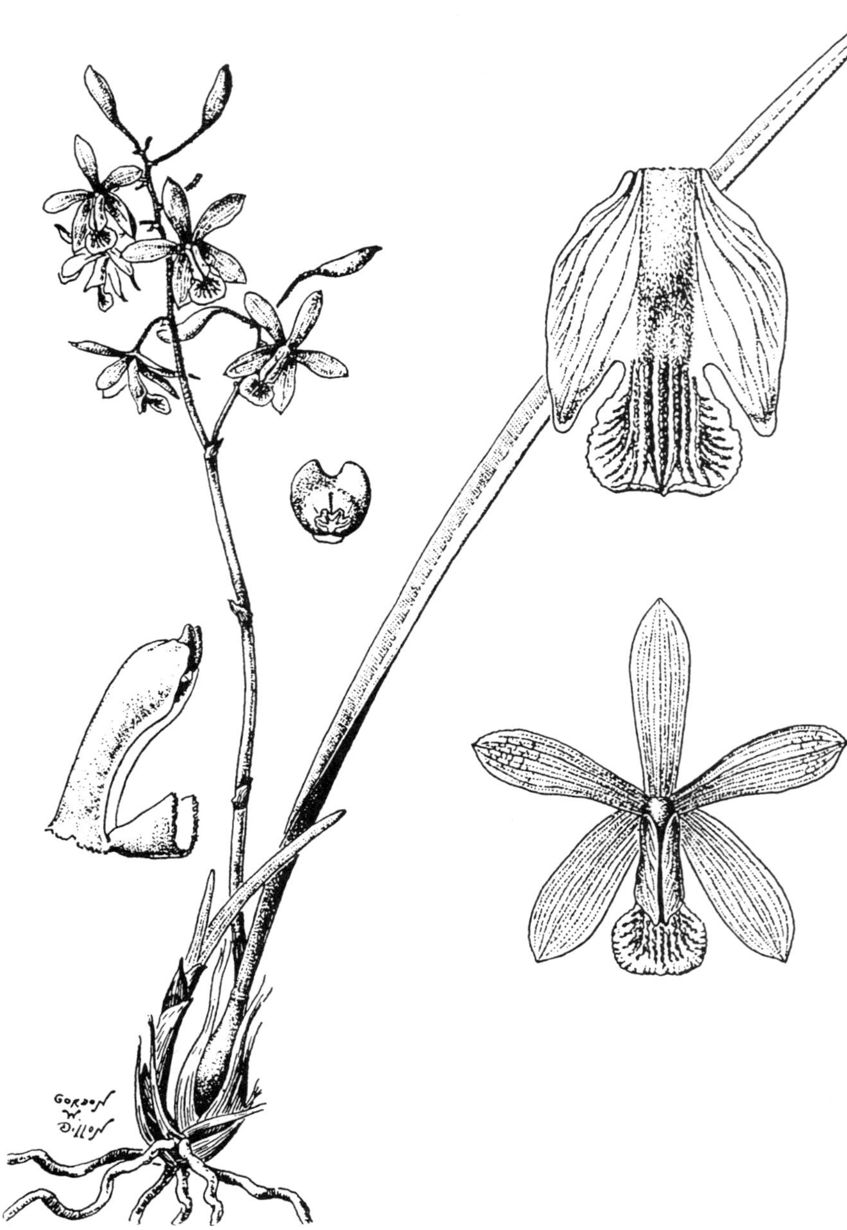

Figure 2-30. *Encyclia pilosa.* Drawing by Gordon W. Dillon. Permission of the *American Orchid Society Bulletin.*

Piracanjuba is a municipality in Goias state. The plants were discovered by an orchid enthusiast named Geraldo Salu and exhibited by Helio de Lima at the local orchid show in Piracanjuba. The plants were collected at 720 m, flowering in December, 1990. They were collected by Menezes, and the type specimen is conserved at the herbarium of the University of Brasilia, UB #10.

Menezes describes the plants as epiphytic, with ovoid, green-brown pseudobulbs to 5 cm tall with one or two leaves that are up to 40 cm long × 2.5 cm wide. The erect inflorescence may be 60 cm high, is unbranched, and has 6–10 flowers. The flowers are slightly fragrant with green-brown sepals and petals and a white lip that may be flushed with a dilute purple and has some central purple veining. Some veining or spotting may also be on the white fleshy forcipate callus ridges. The lateral lobes have brownish veining near their bases with the apical portions stained yellow. The lip is shouldered, has wide lateral lobes that come to rounded, almost truncate tips, and the midlobe is rounded from a broad base with slightly scalloped and turned down edges. The dorsal sepal measures 25 × 7 mm, the petals 24 × 5 mm, but the lip size is not given. The green-yellow column has a white apex, is 12 mm long, and is slightly curved. The anther cap is yellow. Neither auricles nor the ovary surface are mentioned in the description, so must be ascertained when we know more about the species.

Profuse Encyclia

Colombia

(FIGURE 1-5; PLATE 55)

Encyclia profusa (Rolfe) Dressler and Pollard. 1971. *Phytologia* 21:437.

Synonyms

Epidendrum profusum Rolfe. 1914. *Curtis's Bot. Mag.* 140, t. 8551.
Epidendrum oncidioides var. *profusum* (Rolfe) Ames, Hubbard, and Schweinfurth. 1935. *Bot. Mus. Leaflt., Harv. Univ.* 3:107.
Encyclia oncidioides var. *profusa* (Rolfe) Hoehne. 1952. *Arq. Bot. Est. São Paulo,* n.s., 2:154.

Rolfe considered his epithet to indicate a species, *Epidendrum profusum.* When transferred to *Encyclia* by Dressler and Pollard, the taxon was considered a natural hybrid and was written with an × before the epithet *profusa* with the explanation that the parents were *Enc. ambigua* and *Enc. ceratistes.* In using the designation of *Enc. profusa* (Rolfe) Dressler and Pollard, I am returning to the original concept of the taxon as Rolfe described it—in no way a hybrid population—but transferred to the genus *Encyclia.*

This species was apparently born in confusion as a result of a mistaken country of origin. The text with t. 8851 in *Curtis's Botanical Magazine* tells us that the plant was purchased for the Kew collection from Sander and Company, St. Albans, as an example of *Epidendrum candollei.* When it flowered, the plant was immediately recognized as something different. It was suggested that the plant was more closely allied to *Epi. aromaticum* or *Epi. ambiguum,* particularly the latter, both of which are Mesoamerican species. It was never considered that the plant was from a different location,

Colombia, and not México—although the write-up about the new species says, "The locality México is somewhat inferential."

Over the years researchers tried to make this plant fit into what was known about Mesoamerican encyclias, and it finally had its niche as a variety of *Epidendrum oncidioides* following the work of Ames and his co-workers. After 1935 it was more or less abandoned. I too was attempting to figure out this species in my early research with encyclias. Not finding any records in the herbarium of anything like it, I had to keep going back to the original description and plate in *Curtis's Botanical Magazine.* I tried to make things fit, but they refused. At the same time, a single flower from the type specimen of *profusum* had been sent to Ames Herbarium from Kew by the director, Sir Arthur W. Hill. My original notes on this species were dated July 27, 1968. Then one year, while attending the Miami International Show and making the usual round of nurseries, I found a plant in flower that matched the plate in *Curtis's Botanical Magazine* exactly. It was at Redlinger's range, but unfortunately he did not know where the plant had originated. He thought it came from South America. A clue at last! In 1985, attending the show in Medellín, Colombia, I was overjoyed to see cartons of the plants in bloom for sale at the gate outside the show, and at last I knew the country of origin.

The pseudobulbs have two or three leaves, and the panicles of fragrant, wide-open starry flowers exceed the leaves. The sepals and petals are a pale lime green and appear almost transparent. They last as well as other flowers of the genus in spite of their delicate appearance. The lip is white with radiating purple veining, the midlobe dominating and nearly round. The lateral lobes are narrow with rounded tips. The sepals are oblanceolate-oblong and measure about 4–5 mm at their widest and 18–22 mm long. The petals are about the same length and a little wider. The lateral lobes are 6–8 mm long and the midlobe about 10–12 mm across. The total lip length is 16–18 mm, sometimes with a slight indentation at the apex of the midlobe. The column is straight with usual auricles. The ovary is smooth and measures 8–9 mm. The anther cap is bilobed, and each side has a maroon or purple marking that can look like an eye. The whole cap is not "black," however, like those of *Encyclia amicta* or *Enc. thrombodes* nor is the base of the lip pilose. The veining on the midlobe tends to be verruculose and raised, particularly the three to five median veins.

This attractive species could well be used in hybridizing with other encyclias when it is better recognized and identified. It would seem to be endemic to Colombia, though the herbarium record is still poor.

Quesnel's Encyclia

Brazil

(FIGURES 1-5, 2-31; PLATE 56)

Encyclia quesneliana (Hort. *ex* Henshall) Withner. *Comb. et stat. nov.* Basionym: *Epidendrum quesnelianum* Hort. *ex* Henshall. 1845. *The Cult. of Orchids,* p. 113.

Synonyms

Epidendrum roseum Gérard. 1848. *Portefeu. Hort.* 2:266.

Epidendrum jenischianum Reichenbach f. 1853. *Flor. des Serres,* ser. 1, 9:98.
Epidendrum dichromum var. *amabile* Lindley. *Nom. nov.* 1865. *Curtis's Bot. Mag.*, t. 5491.
Epidendrum dichromum var. *amabile* Bateman. *Nom. nov.* 1865. *Curtis's Bot. Mag.*, t. 5491.
Epidendrum dichromum var. *striatum* Reichenbach. 1866. *Gard. Chron.*, p. *218.*
Epidendrum conspicuum Lemaire. 1869. *Illustr. Hort.* 16, t. 592.
Epidendrum amabile Godefroy-Leboeuf. 1887. *L'Orquidophile,* p. 304.
Epidendrum ortgesii Regel. 1891. *Acta Hort. Petróp.* 11(15):473, also 1849, p. 837.
Encyclia amabilis (Linden and Reichenbach f.) Schlechter. 1918. *Beih. Bot. Centralbl.* 36(2):471.
Encyclia ortgesii (Regel) Schlechter. 1920. *Repert. Spec. Nov. Regni Veg.* 7:254.
Encyclia conspicua (Lemaire) Porto and Brade. 1935. *Rodriguésia* 1(2):28.
Encyclia jenischiana (Reichenbach f.) Porto and Brade. 1935. *Rodriguésia* 1(2): 29.
Encyclia ghillanyi Pabst. 1976. *Bradea* 2(14):80.

Misapplications (auctor non)

Epidendrum amabile Linnaeus. 1763. *Species Plantarum* = *Phalaenopsis amabilis* (Linnaeus) Blume *nec* Linden and Reichenbach f. = *Encyclia amabilis* (Linden and Reichenbach f.) Schlechter.

Various pros and cons about this *Encyclia* being a separate species from *Enc. dichroma* have been advanced, and sometimes I have mentally distinguished between the two, and sometimes I have not. After putting together the data necessary to produce Figure 2-31, I was finally convinced that they were two separate species, albeit closely similar. Pabst says in his original description that the main problem with *ghillanyi* is its close affinity with *Enc. dichroma,* and I can vouch for that! He describes it as having uniformly wide, downwardly bent lateral lobes, instead of flaring ones, when the lip is flattened for study. The petals are wider, more spatulate, and the "vivid reddish purple of the flowers with the lighter lip is very impressive." In *Enc. dichroma* the lip is usually darker than the sepals and petals, whatever tone they may have from white to rose or magenta.

Once again, as in the Amicta Complex, we seem to be dealing with a series of relative differences and how they are to be evaluated is difficult to know. Experience, again, comes to the fore. Baron Anton Ghillány has been a figure in Brazilian orchid circles for years and has collected many species, several of them new to the Brazilian flora. His collecting in Bahia brought this species to Pabst's attention, and now we must find out more about Mrs. Quesnel. Please see my remarks under *Enc. dichroma* above and note the variations pictured in Figure 2-31. One can easily conjecture that there has been hybridization in nature between the two species populations, and that further study of the two might be profitable. Has ploidy been a factor, and what relationship might exist between the two?

Please note that the clone named variety *quesneliana* provides priority for an 1843 basionym for the species concept most recently named *Encyclia ghillanyi.* According to our rules of nomenclature, the species epithet must now be changed to *Enc. quesneliana.*

Measurements and shapes of *Encyclia quesneliana* flowers, taken from Pabst's type description of *ghillanyi,* are as follows: dorsal sepal subspatulate with rounded apex, 25 × 7.5 mm, petals spatulate, somewhat clawed,

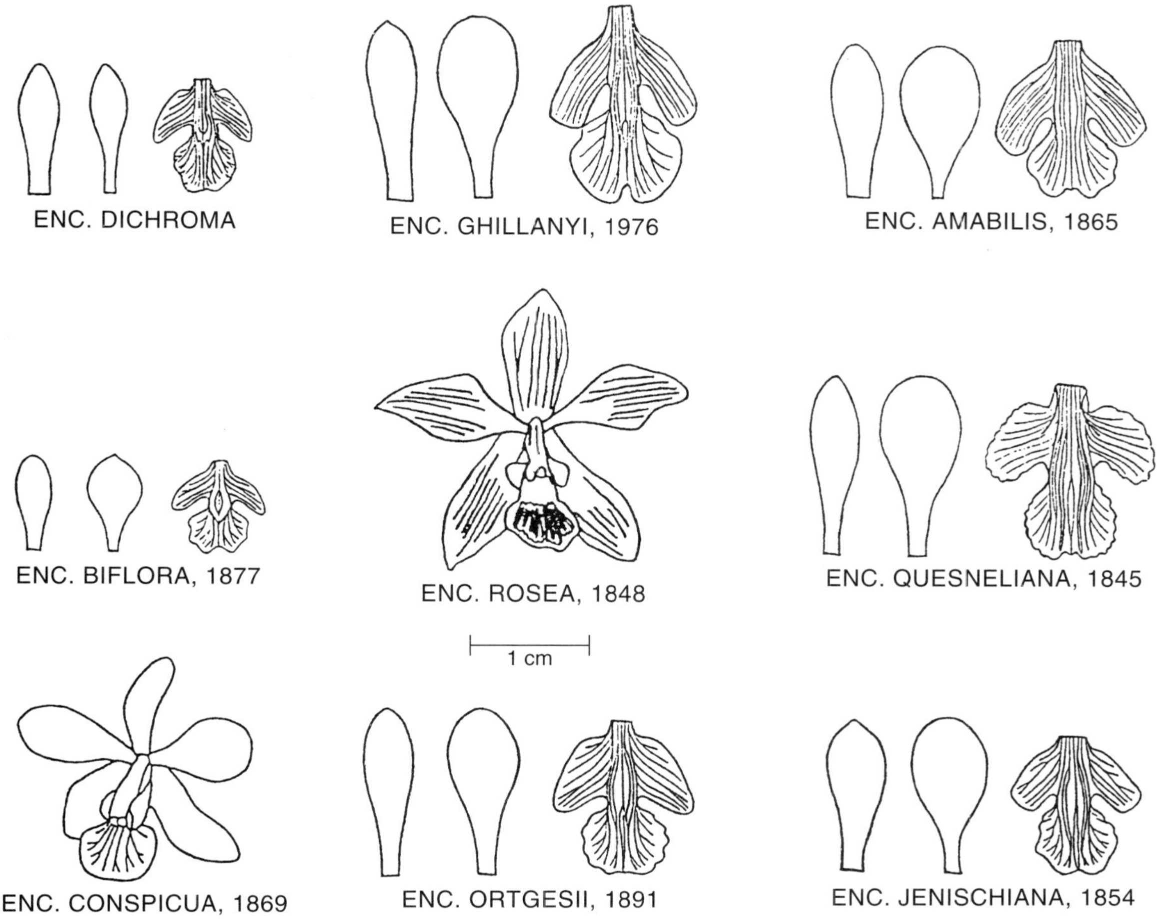

Figure 2-31. *Encyclia quesneliana.* Differences and similarities among *Enc. dichroma* and *Enc. ghillanyi* (now *Enc. quesneliana*) and their varieties. Please see text for details. *Encyclia dichroma* shows an average example; all other drawings are of type specimens conserved at Kew Gardens. Drawings adapted from camera lucida and other studies by Antonio Toscano de Brito.

27 × 13 mm; lip with lateral lobes falcate, oblong, and with broad rounded tips, isthmus short, 4 mm long, and midlobe subrhombic with an indented tip, 3 cm long × about 2 cm wide; ovary 2 cm long and smooth to the touch. The column is distinctly auriculate.

Little more is known about this species, especially in cultivation. It should be grown in baskets of osmunda or on cork slabs hanging in the warm, sunny portions of the greenhouse.

The type description of *Encyclia dichroma* cites a poor horticultural specimen of Quesnel from Havre sent to Lindley. At Kew is a painting (by Quesnel?) of the plant collected in Pernambuco, Brazil, the specimen itself, and one flower from Reichenbach #696. It is noted that the flowers are larger than usual.

Rand's Encyclia

Brazil, Peru?

(FIGURES 1-6, 2-32; PLATES 57, 58)

Encyclia randiana (Linden and Rodigas) Withner. *Comb. nov.* Basionym: *Epidendrum randianum* B. Rodrigues. 1886. *Lindenia* 2:5.

Synonyms

Epidendrum randii B. Rodrigues *ex* L. Linden and Rodigas. 1885–1888. *Vellozia,* ed. 1, 2:123; ed. II, 1891.

Encyclia randii (B. Rodrigues) Porto and Brade. 1935. *Rodriguésia* 1(2):29.

Encyclia randii var. *rondoniensis* Menezes. 1991. *Bol. C.A.O.B.* 3(2):47.

Misapplications (*auctor non*)

Epidendrum atropurpureum var. *randianum* (*randii*) Linden and Rodigas. 1886. *Lindenia* 2:5, t. 49.

Epidendrum atropurpureum var. *randianum* Hort. *ex* Veitch. 1890. *Man. Orch. Pl.,* pt. 6:86.

The name of this species underwent various changes, as the epithets above indicate, but it ended up as *randiana.* What is more important for us, Veitch (1887) considered it a variety of *Encyclia cordigera* (synonym *Epidendrum atropurpureum*). Orchid growers today may still have plants of *Enc. cordigera* var. *randii* in their collections, thinking that it is a color form with reddish veins showing on the lip. This confuses the proper identity of the species and produces an inaccurate geographic range. The epithet *randii* has nothing at all to do with *Enc. cordigera* and should be eliminated from consideration under that species.

Plants of *Encyclia randiana* are among the most elegant of encyclias; they are still comparatively rare in collections, but once seen become necessary! The plants were found originally by Edward S. Rand in two locations that are described in *Lindenia*—one at Teffé, about 2400 km from Pará state, on the Amazon, and a second on the river Tapajós. My first plant, however, came from near Iquitos, Peru, on one of the Amazon tributaries. We still cannot be sure of the actual distribution of the species. In any case it was unusual that such a grand large-flowered orchid could have been missed for so many years when so many other species were being collected, described, and pictured. No doubt it was the bright, warm, humid habitat that made collecting in such areas an uncomfortable ordeal. It does show us, however, that this species likes bright, warm, and humid greenhouse locations, usually suspended in baskets or on cork slabs hanging high toward the glass. It is a summer bloomer.

The species was named by Barbosa Rodrigues to honor Rand, an early director of the museum of Rio de Janeiro and director of the Botanical and Natural History Museum at Manaus, who knew Brazilian orchids well. The large plants can produce panicles of 10–20 or more flowers, even up to 100 says one description from 1889, with branches as long as 60 cm. It sounds somewhat legendary and is something we can look forward to even today. So far as I know, no hybrids have yet been produced, though a selfing or sib crossing would be of first priority, nor have I found a judging record on it. I would rate it as the King of Encyclias, and *Encyclia dichroma* as the Queen, but you will have to make your own choices among the 146 *Encyclia* species from the Bahamas, the Caribbean, México, Central America, and South America.

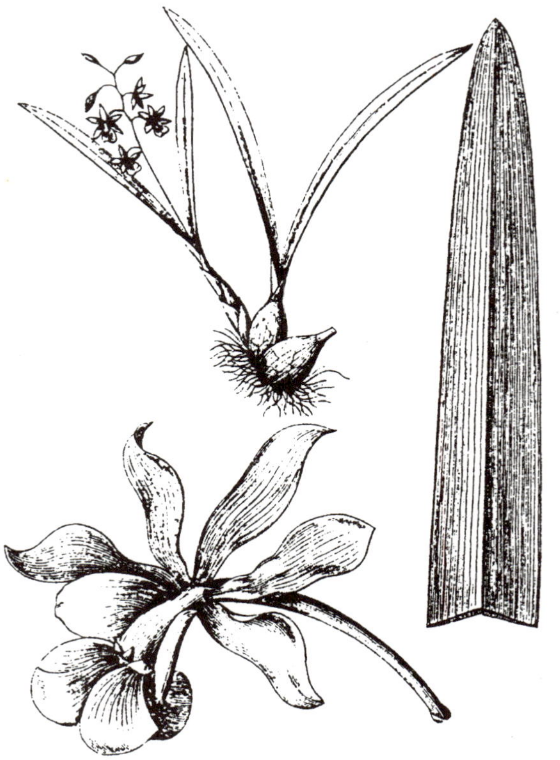

Figure 2-32. *Encyclia randiana.* Type drawing by Barbosa Rodrigues published as *Epidendrum randii* by Cogniaux. Adapted from Plate 16 of *Martius* 3(5).

The large flowers of these *Encyclia randiana* plants have a dorsal sepal measuring 8 mm wide and 32 mm long and are of an ovate-lanceolate shape with a pointed apex. The petals are orbicular-spatulate at the apex and then widely clawed on the basal half, about 5 mm across the claw at its widest, and 13–14 mm across the spatulate pointed tip, 28–30 mm long. They may have crenate ruffled edges, particularly on the terminal portion where it connects with the claw. The lip is deeply lobed measuring also to 28–30 mm long and 23–24 mm across the flattened lateral lobes and about 16–18 mm across the midlobe. The lateral lobes are 13–16 mm long, and their terminal portion is expanded and rounded. The midlobe is faintly kidney-shaped with a lightly notched apex, but it mostly is broadly rounded. The veining spreads out evenly over the disc, emerging from the pronounced bulges of the forcipate callus. The distinct isthmus is 3–5 mm long, and the slightly humped column is around 18 mm. It is somewhat winged, and the wings terminate in a rounded, auricle-like extension, which is not an auricle as such. The ovaries are smooth and about 30 mm long counting the pedicel as well.

The sepals and petals are a deep red-brown, or a sepia color, sometimes with greenish overtones and a thin yellowish border. Lips may have rosy flushes but are generally white with bold red to rosy purple central veining. The bases of the lateral lobes have comparable veins, but their tip portions are clear white.

We will hear more about this exciting species as time goes on. I would predict that it will produce similar but better hybrids than those of *Encyclia cordigera*. These hybrids should not have the problem of curled-in sepal and petal tips, and lip shape and size, flower numbers, and veining patterns should all be pluses. Botanically, I would say that this species is closely related to or part of what I called the Selligera Group or Complex, in Volume V.

The variety *rondoniensis* described by Lou Menezes in 1991 has larger flowers, a pink flush on the disc, and rose-lilac colored veining. It was found at the Orquidário Claretiano in October of 1986, having been collected on the river Claro.

Recurved Lip Encyclia

Brazil?, Guyana?, Venezuela

(FIGURES 1-6, 2-33)

Encyclia recurvata Schlechter. 1919. *Repert. Spec. Nov. Regni Veg.*, Beihefte 6:41.

Synonym

Epidendrum halatum Garay and Dunsterville. 1976. *Venezuelan Orch. Illustr.* 6:126.

Misapplication (*auctor non*)

Epidendrum recurvatum Lindley. 1845. *Bot. Reg.* misc. 54, p. 77.

In the case of this species, two authors named two species with the same species epithet. One species was in the genus *Epidendrum*, the other in *Encyclia*, as cited above. That was no problem. If, however, the *Epidendrum* had needed to be transferred to *Encyclia*, there would have been the problem of two encyclias with the same name—an illegal situation according to our rules of nomenclature. The *Epidendrum* in question, however, is a thin, reed-stemmed orchid with the lip fused to the column. It would never, today, be transferred to *Encyclia*, but, in past years, recognizing this remote possibility, Garay and Dunsterville published the new name of *Epi. halatum* for the Schlechter species.

Carnevali and Ramírez (1988), working with the Venezuelan encyclias, prefer to use the *Encyclia recurvata* of Schlechter, and that is currently the accepted name. Then, to make the problem more complicated, they believe that the plant in question, drawing #772 by Dunsterville, is not *recurvata* nor *halata*, but is a new species they have called *Enc. ivonae* (please see). We therefore must go back to the Schlechter type description of *recurvata* for our information.

The original *recurvata* was imported from Venezuela by Hartmann in Niederhöchstadt, Germany, and flowered in the Botanical Garden in Dahlem, Berlin, where it had been growing since 1912. Its actual origin in Venezuela was not known. Schlechter said in his description that the mid-sized flowers were like those of *Encyclia oncidioides*, but that they were smaller, had smaller lateral lobes, and a clearly clawed midlobe. The flowers had a good fragrance and were characterized especially by the backwardly reflexed sepals. The flower panicles reached just above the leaves, about 20 cm long × 8 cm wide. The sepals and petals were green suffused with brown, and the labellum was yellow-white with dark red veins. The column had auricles, and the ovary was smooth. The shapes and measures were as follows: pseudobulbs 4–5 cm tall with two leaves to 30 cm long × 2.5 cm wide; dorsal sepal oblong, bluntly pointed 16 mm long; petals obliquely oblong-spatulate, curved slightly forward; labellum 15 mm long, and lateral lobes on flattened lip 17 mm; lateral lobes falcate-ligulate; midlobe orbicular, and apiculate, 8 mm across with thickened veins from the callus decurrent on the disc, particularly the central three. The column was 7 mm; and the pedicellate ovary 17 mm. The Schlechter species was illustrated in Mansfeld's 1929 *Atlas der Kordillerenstaaten*, t. 7, no. 24 (Figure 2-33).

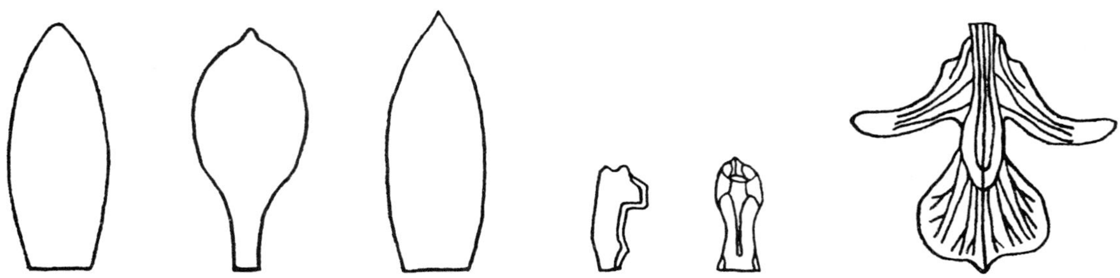

Figure 2-33. *Encyclia recurvata.* Drawing of the type specimen destroyed in Berlin. Adapted from Mansfeld's *Atlas der Kordillerenstaaten*, 1929.

Remote-flowered Encyclia

Venezuela

(FIGURES 1-6, 2-34)

Encyclia remotiflora (Schweinfurth) Carnevali and Ramírez. 1988. *XLVI Exposicíon de Orquídeas, Comité de Orquideología,* Bol. No. 23:75.

Synonym

Epidendrum remotiflorum Schweinfurth. 1962. *Bot. Mus. Leaflt., Harv. Univ.* 20(1):18, pl. 7.

The plants of this species are short, usually with single leaves, sometimes two, and the racemes stretch above the leaves but only produce a few distant flowers. The drawing by Elmer Smith shows its habit clearly (Figure 2-34). The type specimen was collected on the right-hand bank of the Rio Orinoco in the Amazonas of Venezuela by Wurdack and Adderly, #43155, in June, 1959. It is conserved in the herbarium at the New York Botanical Garden. The herbarium notes tell us that the sepals and petals are old gold in color, the lip white at the tip and rose-streaked on the disc. The ovary surface is smooth. Carnevali et al. (1994) say it is found only in that area and seems to be endemic there. The species would seem to be closely related to *Encyclia pachyantha* and *Enc. latipetala* (please see), but differs in

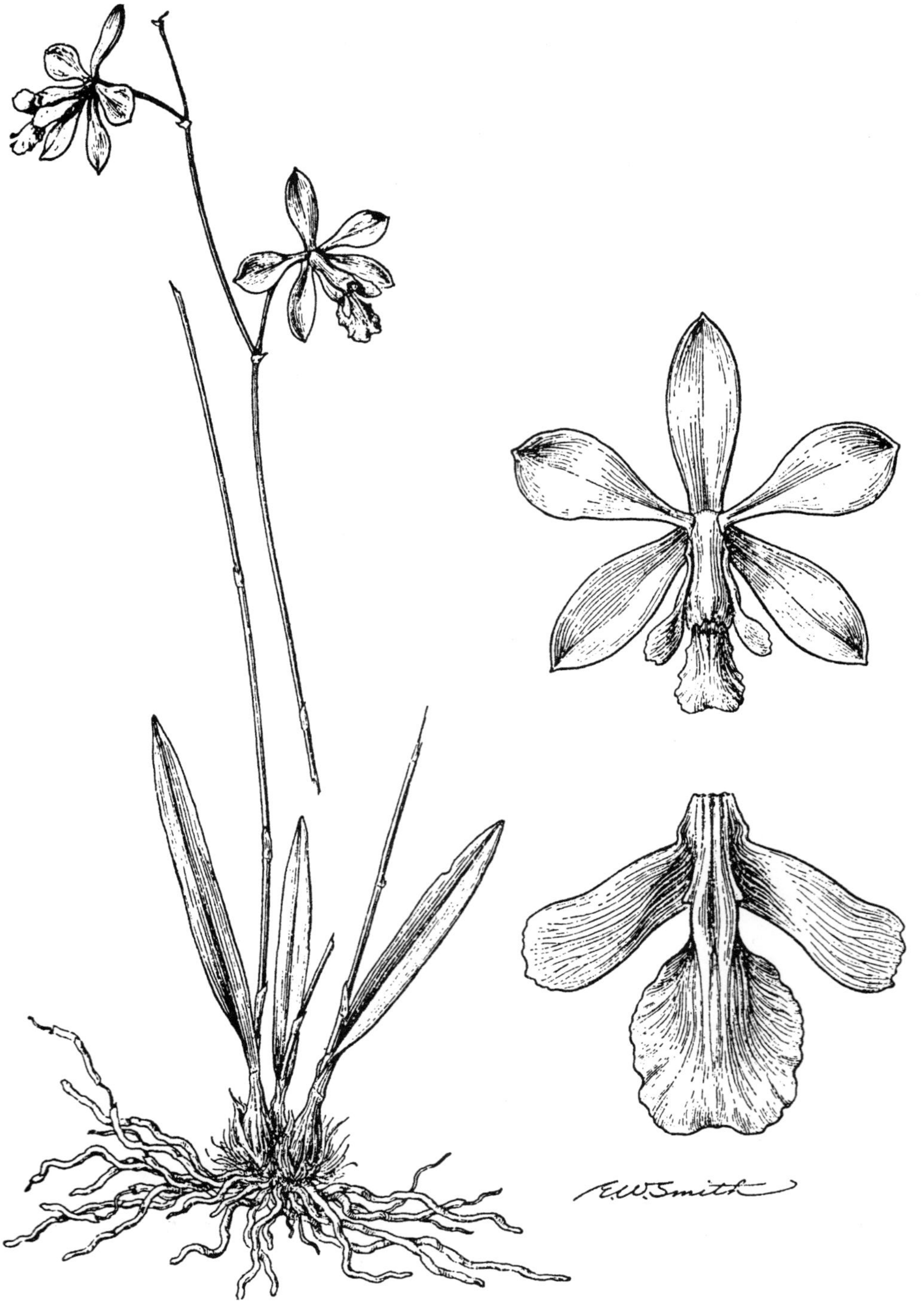

Figure 2-34. *Encyclia remotiflora*. Drawing of *Epidendrum remotiflorum* by Elmer W. Smith from the *American Orchid Society Bulletin*. Permission of the American Orchid Society.

several small ways from either species—for instance the narrower, usually single leaf and the smaller pseudobulb. The leaves grow to 8.8 cm long × 1.2 cm wide with pseudobulbs to 2 cm long. The sepals are oblong-lanceolate, pointed, and about 18–19 × 6–7 mm. The petals are spatulate with rounded tops and reach 18 × 9 mm. The lip has a wedge-shaped base with oblongish lateral lobes that end in truncate and serrate apices. The midlobe is obovate to suborbicular with raised veins. The column is about 18–19 mm long and has prominent auricles. The ovary surface is smooth.

Little else is known about this species, and it does not seem to have been in cultivation. The flowers turn brown on drying, not black.

Folded-lip Encyclia

Colombia, Ecuador?

(FIGURES 1-6, 2-35; PLATE 59)

Encyclia replicata (Lindley and Paxton) Schlechter. 1920. *Repert. Spec. Nov. Regni Veg.* Beihefte 7:255.

Synonym

Epidendrum replicatum Lindley and Paxton. 1851. *Paxton's Flow. Gard.* 2:167, Fig. 224. Also see *Paxton's Flow. Gard.* 3:38 where it is published again in 1884 with the figure numbered 210, but with similar text. Also, the same illustration and account are in *Le Jardin Fleuriste* for 1852, 2:121. A different drawing but similar text is also published in *The Garden Companion and Florists' Guide,* part 5, May, 1852.

The species was apparently imported by Linden and grown and displayed by Sigismond Rucker Esq. It won a silver Knightian Medal at the July meeting of the Royal Horticultural Society in 1851. A painting of the flowers dated July, 1851, is at Kew. Reproduced here as Figure 2-35, this painting is likely the work of Lindley. The text from *Le Jardin Fleuriste* adds the detail that the original plant was collected by Funck and Schlim for Linden.

The large nodding panicles had yellow-brown flowers toned with darker chocolatey brown. The sepals and petals were bordered in yellow, and the white lip was veined in rose. The sides of the lip are folded or reflexed downward along the midvein until their backs almost touch. The panicles can reach more than a meter long, and a specimen plant with several panicles was a sight to see. In the herbaria I have found several specimens mislabeled *Encyclia replicata* that were *Enc. howardii,* a Cuban species not occurring in South America, but I have not found any Colombian specimens that match the description. In the plant on the left side of Figure 2-35, the lip detail is shown with a sharp, apiculate midlobe that is replicate. It seems to be lacking in the plant on the right side of that figure, and I sometimes wonder if the two drawings are really of the same plant.

The leaves can reach 51 cm long × 3.6 cm wide, and the pseudobulbs can be 10 cm tall. The sepals are oblong-lanceolate, pointed, to 20 × 6 mm. The petals are more spatulate, oval-orbiculate with a claw, abruptly pointed, to 19 × 6.5 mm. Lips are 16 mm long, slightly clawed, and have obovate lateral lobes with rounded apices. The midlobe is somewhat triangular, about 7 mm broad. The column is auricled, and the ovary surface

Figure 2-35. *Encyclia replicata.* (*Left*) Illustration accompanying the original publication of the species as *Epidendrum replicatum* by Lindley and Paxton in 1851 (in *Paxton's Flower Garden,* figure 224). The illustration and information were published again in 1852 (in *Le Jardin Fleuriste*) and 1884 (in *Paxton's Flower Garden).* (*Right*) A different illustration but similar text as for the plant on the left was published in *The Garden Companion and Florists' Guide* of 1852. The plant with its long (1.2 m) panicles, apparently made a lasting and exciting impression.

is smooth. The description by Lindley in Paxton concludes, "The lip, which is white, streaked with pink, is remarkable for the manner in which the two sides are turned downwards, so that their backs actually touch. Most nearly allied to *Epidendrum diotum*, a native of the same country."

I have thought that *Epidendrum replicatum* as described in Paxton is now *Encyclia selligera.* No Colombian specimen I have seen in any herbarium looks like the portrait in Paxton, fig. 224, but that of *Enc. selligera* from Central America (Vol. V) matches it reasonably well. In the write-up of the picture it says, "We *presume* it to be one of Mr. Linden's importations." The watercolor at Kew marked "New Grenada, Linden, M. Rucker, July, 1851" is also a possible match with *Enc. selligera*. Did the idea of it being a Colombian specimen result from this notation of New Granada? Most likely so. But here we are today trying to fit it into Colombian orchid accounts that have, so far, never been a match for it.

Another drawing of *Epidendrum replicatum* exists: the same Rucker plant from Wandsworth was exhibited at one of the Chiswick Fetes, and it is pictured in *The Garden Companion* as cited above. Also, there was an idea that the species could be Cuban—at least one Wright specimen is identified as *Epi. replicatum.* This specimen may also be what is now called *Enc. insidiosa.* With this aura of confusion about the identity of this species, even with the various publications about it, it could be listed as a dubious taxon. I have found it confusing, especially, as I have not had any herbarium material that I could positively identify with this species.

Santos Dumont Encyclia

Brazil

(FIGURE 1-6; PLATE 60)

Encyclia santos-dumontii Menezes. 1991. *Bol. C.A.O.B.* 4(1):12.

Antonio Prenholato, an orchid enthusiast from the Brasilia Orchid Society and an Air Force captain, discovered this species while on assignment to an Air Force base and military reserve near Caximbo Hill in Brasilia. Since the region is protected, the reserve's flora and fauna have not been much disturbed. The plants grew epiphytically in a flooded wilderness area. Because it was discovered on the birthday of Alberto Santos-Dumont, famous aeronaut and the Father of Aviation in Brazil, the species was named in his honor.

The type description says that the orchids had characteristics unusual for the genus. The pseudobulbs were small and had a fusiform shape. The leaves, up to three in number, were leathery, arched, 34 cm long × 0.9 cm wide, and they had acute apices. A few flowers grew on a short stalk not higher than the leaves. The column was large for the size of the flower and had a prominent rostellum. The sepals were green-brown, 17 × 5 mm, and the petals the same green-brown color and spatulate, 15 × 5 mm. The lip was small, and the midlobe rounded, white with purple veins. It had some spotting near the base, and the midlobe had a wavy edge. The column measured 10 mm.

This species needs to be grown from seed so that we may begin to know more about it beyond a single type specimen preserved in a Brazilian herbarium. So many encyclias seem to be in this same category—they have been found only once, or a few times at best, and would therefore be only of a local or endemic distribution. In Brazil, especially, so many of these locally found species are becoming apparent that one is challenged to provide some explanation beyond the great range of environmental niches that such a large tropical country as Brazil does provide. Size alone, however, cannot account for the speciation that has occurred there during many millennial years since Brazil split off from Africa as the result of continental drift.

Schmidt's Encyclia

Brazil

(FIGURE 1-6; PLATE 61)

Encyclia schmidtii Menezes *ex* Fowlie. 1991. *Orchid Digest* 55(2):92.

This species, as a result of the renewed interest and fieldwork on the Brazilian species by Fowlie, Duveen, and Menezes, is named in appreciation for the role of Antonio Schmidt, an orchid enthusiast, who helped in the Brazilian research. The plants were found by Schmidt in shrubby prairie around Sorriso next to Chapada dos Parecis in Mato Grosso state. The plants were flowering from September through October, the Brazilian spring. The type specimen is conserved at the herbarium of the University of Brasilia, UB #06.

The sepals and petals are a greenish brown with twisty curly edges, the green more pronounced toward their bases. They are oblanceolate to elliptic in shape, the wide green bases about a third of their total length. The lateral lobes have wide shouldered bases with stubby quadrangular shapes and truncate apices. The midlobe is rounded and a little retuse at the apex. The lateral lobes are green-brown at the base with maroon veining, and greenish yellow for their upper third. The midlobe is white with seven verrucose maroon-purple veinings over the disc. The column is green at its base and has a rosy suffusion toward its apex. At the end of the wingy edges are teeth. The eight flowers on the type specimen were in a raceme 45 cm long and were slightly fragrant.

Measurements in the type description include pseudobulbs 3.5 cm wide, two leaves about 18 cm long × 0.9 cm wide; sepals 22 × 6 mm; and petals 21 × 6 mm. Lip measurements were not given, nor was there mention of any auricles or teeth on the column. Plate 61, however, shows what appear to be small toothlike projections. The veins over the callus ridges are also colored except at the very base of the lip where there is a short claw.

Stiff-stalked Encyclia

Peru

(FIGURES 1-6, 2-36; PLATE 62)

Encyclia sclerocladia (Lindley *ex* Reichenbach f.) Hoehne. 1952. *Arq. Bot. Est. São Paulo* 2(6):155.

Synonym

Epidendrum sclerocladium Lindley *ex* Reichenbach f. 1856. *Bonplandia* 4:326.

This species has at present a somewhat controversial status with a debatable distribution. The type specimen was collected by Warscewicz, #48, and is presently in the Reichenbach Herbarium in Vienna with sheet #925, and it includes a sketch by Reichenbach f. of flower details. It was collected on the river Marañon in Peru. The specimen shows a stiff, large-diametered flower stalk with definitely small flowers. Garay's drawing of a boiled-up flower from the type is found in Figure 2-36.

Lindley's type description states that the ovaries of the flowers are rough and asperous as are the branches of the stalks that make up the panicle. Lindley did not have the leaves or pseudobulbs available for description. His dimensions and description of the flower include the following details: ovary and pedicel 17–18 mm, triangular, rather arcuate; dorsal sepal 12 mm long × 3.5 mm wide, rigid, fleshy, five-nerved, cuneate-oblong, pointed; petals flat, fleshy, five-nerved, long attenuate below, cuneate-lanceolate, pointed, 12 × 2.5–3.0 mm; lip about 1 cm long, somewhat shorter than the lateral sepals, lateral lobes lightly three-nerved, triangular-lanceolate, pointed and incurved, terminal lobe lightly multinerved, broadly ovate, abruptly acute, three-grooved callus on the isthmus extending to the apex and lateral veins somewhat thickened; column 6 mm long, widening toward the tip and winged, the wings rounded and incurved toward the stigma. No auricles are mentioned. The dissection drawing shows these details well.

Notes on a specimen by Hutchinson, #1523, collected in Amazonas Department, Bagua Province, Peru, on the river Marañon, tell us the sepals and petals are green with maroon veining. The lateral lobes of the lip are green, and the midlobe is yellow with red veining.

The debate comes in when considering *Encyclia bradfordii* from Venezuela. *Epidendrum bradfordii* (see Vol. IV, also see above) was originally described in 1864 by Grisebach from Trinidad and Tobago, and in 1856 *Epi. sclerocladium* was published by Lindley. In 1986 Carnevali and Ramírez transferred the *bradfordii* to *Encyclia,* and in 1988 made *sclerocladia* a synonym under *bradfordii* while identifying one of Dunsterville's collections—from the coastal mountains of Venezuela at about 800 m elevation —that had not been previously published. From my perspective, certain differences between the two species prevent making them synonymous, in particular their lip outlines. The illustration of *bradfordii* (see Vol. IV, p. 40) adapted from a drawing by Reichenbach of Dr. Bradford's 1846 collection and the drawing of the type specimen lip outline in the present volume (see Fig. 1-11) are different. The asperous ovary and flower stalk of *sclerocladia* are not apparent in the *bradfordii,* and the Bradford specimen has large and distinct auricles instead of just a winged column with the wings folding in at the apex. I therefore feel they should be kept as separate entities until we know a little more about the two species. I believe that the good drawing #401 in Volume I of Dunsterville and Garay's (1979) *Field Guide* should also be called *Enc. bradfordii,* reserving the *Enc. sclerocladia* for a species so far known only from Peru with no known intermediate collections that I have been able to discover.

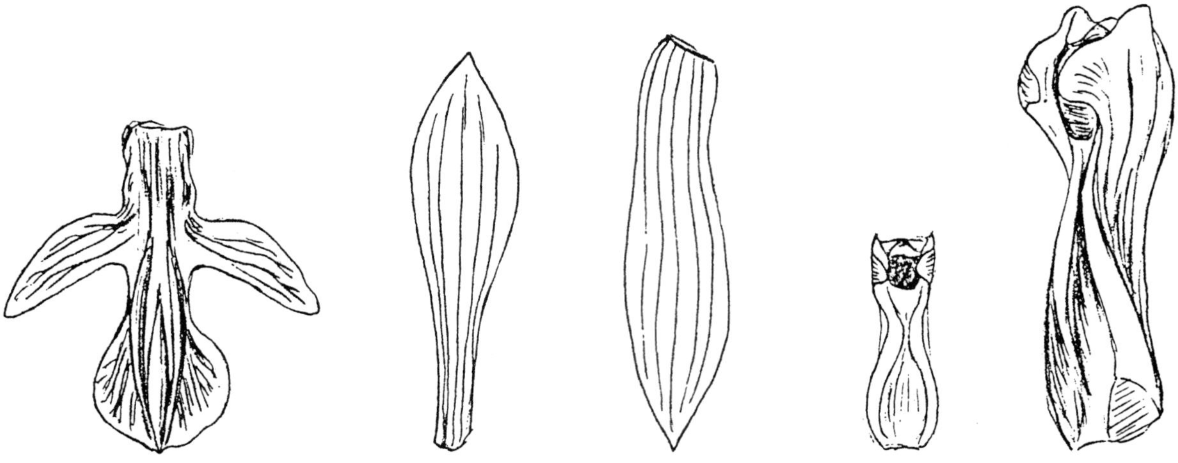

Figure 2-36. *Encyclia sclerocladia.* A flower from the type specimen of *Epidendrum sclerocladium,* Warscewicz, *s.n.,* Peru, conserved in the Lindley Herbarium at Kew Gardens. Note the column auricles, also the callus keels that meet at the apex of the lip. Camera lucida drawing by Leslie Garay.

Seidel's Encyclia

Brazil

(FIGURE 1-6; PLATE 63)

Encyclia seidelii Pabst. 1976. *Bradea* 2(14):81.

The Seidel family has been in the orchid nursery business in southern Brazil for many years, and several new orchids have been found through their efforts. It is fitting that one of the new *Encyclia* species be named in their honor, the type specimen having been found by Alvim Seidel, #1111, in Minas Gerais near Datas, and flowered at their nursery in October, 1974.

The inflorescence is a tall simple raceme, to 70 cm, with 8–10 distant flowers toward the upper portion. The sepals and petals are greenish red-brown and the lip has lateral lobes that are greenish with brown veining and a midlobe of rosy magenta. The sepals are lanciform, obtuse, 17–18 × 5 mm, and the petals are subspatulate, obtuse, 17 × 4.0–4.5 mm across the widest part. The lateral lobes are distinctively obtusely rounded and small, about 7 mm long, not much larger than half the midlobe, which is broadly elliptic and obtuse. and without an apiculus. The central veins extend to a point at the apex. The labellum is slightly clawed, 17 mm long, and has a 2 mm isthmus. The midlobe is 8–9 mm long × 7 mm wide. The column is 8–9 mm long, has small auricles, and is obscurely sigmoid with a pink-rose anther cap that is distinctly furrowed between the two halves. The plants are usually two-leaved.

This attractive, unique species is readily distinguished though we know little about it.

Espírito Santo Encyclia

Brazil

(FIGURE 1-6; PLATE 64)

Encyclia spiritusanctensis Menezes. 1991. *Orchid Digest* 55(1):23.

Synonyms

Encyclia megalantha var. *spiritusanctensis* Menezes. 1990. *Inst. Bras. Meio Amb. Comunicado Téch.*, p. 1. *Nomen nudum.*

Encyclia spiritusanctensis Menezes. 1990. *Bol. C.A.O.B.* 2(4):7. *Nomen nudum.* A revised article follows in 1991, 3(1):16 validating the species with the type data.

This species was first published in 1990 as a *nomen nudum,* an oversight leaving out the citation for the type specimen and its location in an herbarium. The taxon was republished in 1991, as cited above, with all the pertinent information intact. The large flowers are charming with their rosy pink color and darker purplish veinings. The expanded tips of the lip lateral lobes, as well as the midlobe, have a wavy crenate outline, which, along with the flower size, color, and fragrance, will give this species considerable horticultural merit. But first, we should grow new plants from seed to foresee any such demands. The plants were discovered growing in full sun on rock outcrops, flowering in February and March, the Brazilian summer.

The flowers vary from 3.0 to 3.5 cm across with cinnamon brown sepals and petals. The dorsal sepals reach 23 × 8 mm and are obovate-lanceolate in shape with obtuse tips. The lateral sepals are shorter, but about the same width. The petals are spatulate-clawed, 18 × 12 mm and obtuse. Both sepals and petals are greenish white toward their bases. Lip measurements are not given. The lip lateral lobes are broadly strap-shaped, rounded, and expanded at their tips, and poorect along the column. The column is 18 mm long, white, and has a pair of "forward-pointing, forward-projecting acuminate auricles." The scape is less than 1 m long.

Pseudobulbs may reach 10 cm long × 3 cm wide, and they may have three or even four leaves that can measure 35 cm long × 3.5 cm wide. The panicles can be a meter or longer with as many as 40 flowers. The plants were found by Hermann Kundergraber in 1990, and the type specimen is conserved at the herbarium of the University of Brasilia. There is a distinct resemblance of the flowers to those of *Encyclia osmantha,* but until there is better knowledge about variations between the two species (flower size, flower color, lip details, column structure, and geography), they must be considered separately.

Steinbach's Encyclia

Bolivia

(FIGURE 1-6; PLATE 65)

Encyclia steinbachii Schlechter. 1922. *Repert. Spec. Nov. Regni Veg.*, Beihefte 10:47. Also see Mansfeld, 1929, *Atlas der Kordillerenstaaten*, t. 139, no. 544.

Synonym

Encyclia buchtienii Schlechter. 1929. *Repert. Spec. Nov. Regni Veg.* 27:62. (Note: This species is different from *Epidendrum buchtienii*, which is based upon Buchtien #2565.)

This species may be the most southern *Encyclia* species in South America. Very little is known about it besides the type description. Fred Fuchs in Naranja, Florida, once had a few plants in his collection that produced nodding panicles of large yellow-green flowers with white lips and purple veining. Each had two or three leaves, and the pseudobulbs tended to be long and narrow. The type specimen was collected by J. Steinbach in December, 1916, in the Bolivian forests of Buenavista at about 450 m elevation.

The sepals of *Encyclia steinbachii* are described as oblong, obtuse, about 18 mm long, with the bases a little narrowed. Petals are oval-spatulate, obtuse, equal to the sepals in length. The lip is 17 mm long with a short claw. When flattened the distance between the enlarged tips of the lateral lobes is also 17 mm. The apices are clear of veining. Lateral lobes are narrowly oblique, oblong, and rounded, and measure about 10 mm. Three slightly thickened veins extend from the callus toward the apex. The midlobe is rounded and about 10 mm long and broad, slightly notched at the tip. The lip has a green base and is otherwise white with deep purple veining. The column measures 10 mm and has auricles that are narrow and pointy. The ovary surface is a little rough to the touch. My card specimen samples flowered in either July or August in cultivation.

Very likely *Encyclia buchtienii* is a synonym of this species, but until more fieldwork or collected specimens can clinch the identifications, we cannot be certain. It is based on Buchtien #4211 and was probably destroyed in Berlin. It was collected at Milluguaya, La Paz, in Nord-Yungas at 900 m and flowered in December, 1917. Schlechter stated it looked similar to *Enc. ambigua or Enc. trachychila* from Central America. The type description of *Enc. buchtienii* shows the pseudobulbs were cylindrical, 2.5–3.0 cm wide, and had three obtuse, leathery leaves 50–60 cm long × 2.3–3.2 cm wide. The inflorescence was paniculate, more than a meter long, nodding, and loosely flowered. The sepals were oblong, pointed, and 13 mm long, and the petals were obovate-spatulate, obtuse, about as long as the sepals. The lip had a short claw and a broadly oval midlobe slightly heart-shaped at the isthmus and about 10 mm long. The lateral lobes were oblique, obtuse with a 5 mm interior length. The midlobe was subreniform-orbicular with an undulate margin and an indented apex. The column measured 7 mm and had auricles. The pedicellate ovary was smooth, 20–23 mm long.

Starry Alanje Encyclia

Colombia, Costa Rica, Panama, Venezuela

(FIGURES 1-6, 2-37)

Encyclia stellata (Lindley) Schlechter. 1914. *Die Orchideen,* ed. 1, p. 214.

Synonyms

Epidendrum stellatum Lindley. 1853. *Fol. Orch. Epid.,* p. 21.
Encyclia hunteriana Schlechter. 1922. *Repert. Spec. Nov. Regni Veg.,* Beihefte 17:46.
Epidendrum alanjense Ames. 1922. *Sched. Orch.* 1:13.
Encyclia powellii Schlechter. 1922. *Repert. Spec. Nov.,* Beihefte 17:46. Please see Volume V.
Encyclia alanjensis (Ames) Carnevali and Romero. 1994. *Lindleyana* 9(1):67.

From the start there seems to have been some debate over the nature of this species and what it should be named. Lindley's name for it is based on an unnumbered Purdie specimen collected in Colombia and now preserved at Kew. Garay, however, has stated that he finds this species to be identical to *Cymbidium glandulosum,* which was collected by Humboldt, Bonpland, and Kunth and based on a plant from Mount Avila north of Caracas in Venezuela. Researchers today find that the *C. glandulosum* type is a better match for what is called *Encyclia diurna* (please see), and Carnevali et al. (1994) state that *diurna* is one of only two species of *Encyclia* found in that part of Venezuela. *Encyclia diurna* is primarily a Venezuelan species and does not seem to be reported from Colombia. It should not now be considered with *Enc. stellata.* The Panamanian–Costa Rican species *Enc. alanjensis,* however, seems possibly to be a match, as Carnevali and Romero have pointed out. This provides one more species with distribution in both Central and South America—along with *Enc. cordigera, Enc. ceratistes,* and possibly *Enc. alata* (Escobar, 1991,1994).

Encyclia stellata is another of the green-brown encyclias with a white lip that are collectively confusing and have problems of recognition. This starry species has a rounded to oval and somewhat pointed midlobe on the lip and generally lacks purple veining that is so common on most of these green-brown species. The midlobe shows five main raised veins extending from the callus over the disc, the three central raised veins most prominent. Under the microscope the veins are minutely hairy or papillose. Also, rather distinctively, the column has auricles as such, but the wingy or angulate edges (as they have been called) of the column dip down to form a small point or tooth on either side of the stigmatic area. The pedicellate ovary is scabrous. The lateral lobes of the lip are triangular and slope down from a shouldered base to obtuse tips. On most specimens I have measured the lip extends twice the length of the 6.5–7.0 mm long column. The anther cap is white. The dorsal sepal is oblanceolate (Ames says elliptic-oblong) and measures 16 mm long. The petals also are oblanceolate, to a greater degree, about 15 × 3 mm, and the lip is about 13 mm long with the midlobe 8 mm long × 7.5 mm wide. The flowers are either in short racemes or in short panicles with the plants growing in dry or seasonally dry forests in western Venezuela, according to Carnevali et al. (1994). The plants seem little known in Costa Rica or Panama, and there are no records of the species in cultivation.

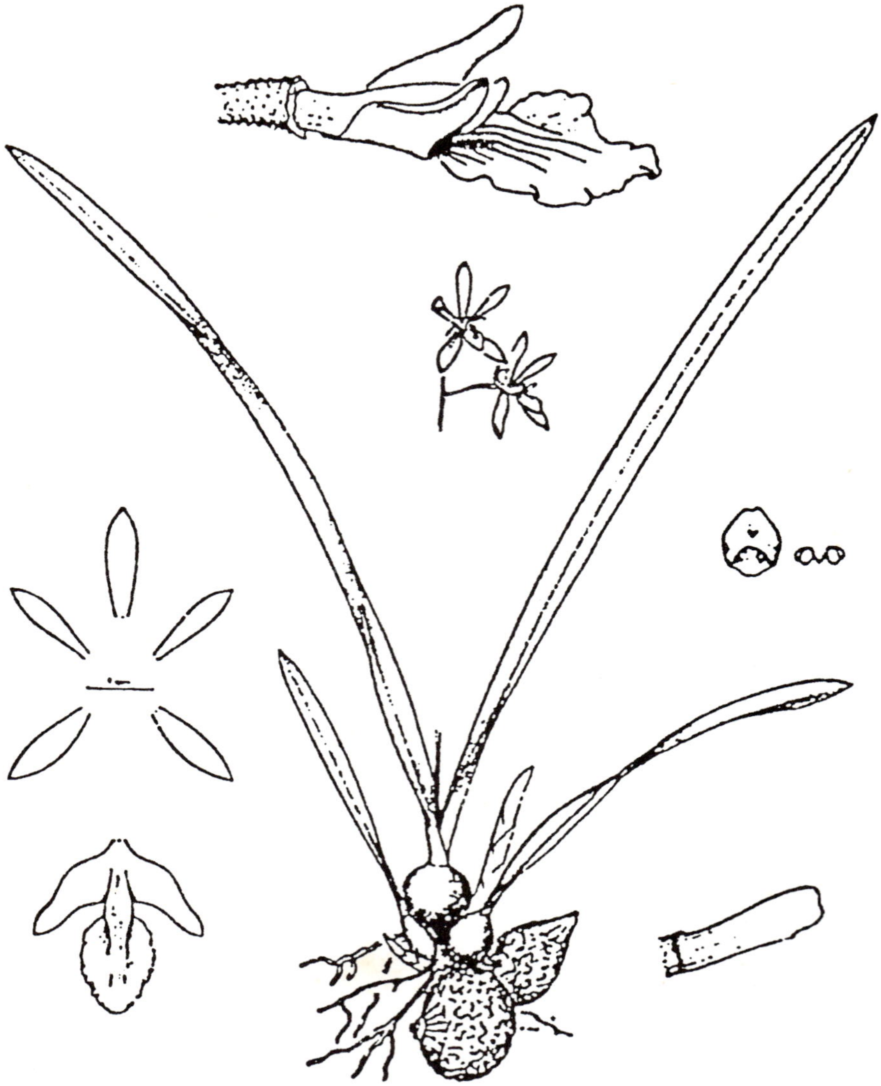

Figure 2-37. *Encyclia stellata.* This species is mostly found in Colombia, Panama, and Costa Rica, but Dunsterville thought he had found the species in Venezuela. I am inclined to identify his drawing #940 as *Encyclia picta* (please see Figure 2-29) rather than *Enc. stellata.* Please see text for both species; also see *Enc. stellata* in Volume V.

Taruma Encyclia

Brazil

(FIGURE 1-6; PLATE 66)

Encyclia tarumana Schlechter. 1925. *Beih. Bot. Centralbl.* 42(2):106.

Synonym

Epidendrum tarumanum (Schlechter) Hawkes. 1956. *Orquídea* 18:178.

Taruma is near Campinas in the Amazonas state of Brazil. This vast state has several major rivers and innumerable smaller ones; most habitation is along the waterways. The region abuts Colombia on the north and Peru on the west. The type specimen was collected by G. Huebner, #188, flowering in September. The sepals and petals were bright green at their bases with lilac veinings towards the tips, and the lip was white with bright lilac veins at the base. The species has not been common in cultivation and little beyond the type description is available about it.

The plants are epiphytic and reach 60 cm high. They make large clumps. The pseudobulbs are about 3 cm apart, separated by a rhizome of 5–6 mm in diameter. They are narrowly pear-shaped, to 4 cm tall and 1.5 cm in diameter, and bifoliate. The leaves are erect, narrowly linear, subacute, leathery, and to 38 cm long × 8–9 mm wide. The flower stalk is graceful, has a few branches, and 10–17 flowers. The stalk is smooth and about 25 cm long. The sepals are narrow and about 2 cm long. The petals are oblanceolate, bluntly tipped, narrow at the base, and measure 1.85 cm long. The lip is 1.85 cm long. The lateral lobes are bluntly triangular and oblique, with apices about 1.2 cm apart. The midlobe is circular, rounded or obtuse at the tip, with undulating margin. It is 1 cm long × 8 mm broad. The column is 1.2 cm long with auricles, incurved, and somewhat narrow. The pedicellate ovary is smooth, about 2.3 cm long.

This desirable species should also be raised from seed whenever a plant becomes available for pollination. Who will specialize in seedlings of this large genus of desirable species?

Pflanz's Encyclia

Bolivia, Brazil?, Colombia?, Peru

(FIGURE 1-7; PLATE 67)

Encyclia thrombodes (Reichenbach f.) Schlechter. 1921. *Repert. Spec. Nov. Regni Veg.*, Beihefte 9:154.

Synonyms

Epidendrum thrombodes Reichenbach f. 1877. *Linnaea* 41:79.
Encyclia pflanzii Schlechter. 1922. *Repert. Spec. Nov. Regni Veg.*, Beihefte 10: 46. Also see Mansfeld, 1929, *Atlas der Kordillerenstaaten*, t. 543.
Epidendrum pflanzii (Schlechter) Hawkes. 1956. *Orquídea* 18:177.

The records on *Encyclia thrombodes* named by Reichenbach f. were made from a Peruvian plant grown at the Veitch Nursery. Flowers were collected originally by C. Pflanz in 1913 near Villa Montes in Bolivia. Recently two different hobby growers had sent me flowers for identification. This in turn led me to search the records again for a proper identification. As a result, I had to consider the names centered about the Brazilian species called *Encyclia bicornuta,* for it is easy to recognize the column without auricles, the deep maroon anther cap, as well as other characteristics of that species. Where do the characters start and stop? To what degree are species distinct from one another—or do they represent more than usual variation in a single species? These are questions without distinct answers, especially when we are dealing with a group or complex of closely related entities. Shall I be a lumper or a splitter?

In this case I decided on splitting the complex into *Encyclia thrombodes,* then *Enc. amicta* and *Enc. conchaechila.* If you prefer to consider them all as one taxon, then use the epithet of *Enc. amicta,* as that name has overall priority in the complex (please see above). Their differences are comparatively relative rather than absolute. Again, it would be nice if we had a DNA profile of the various clones to compare with one another. Do you believe that such a profile could answer these questions about species limits? No doubt it will sometime be possible, but meanwhile we must mostly rely on the comparative anatomy of details to make possible distinctions. Also, please see the description and illustration of *Enc. pilosa* from Peru. That species has been a questionable synonym of this species.

At this point I believe that I can distinguish flowers of *Encyclia thrombodes* by a slightly larger size, their longer than broad lip midlobes with ruffled edges and obtusely pointed tips, their reflexed sepals and petals, the very rounded shoulders of the lateral lobes, the parallel outer edges of the lateral lobes, and the occurrence of the plants in Bolivia and Peru. The open flower shows the upper edges of the lateral lobes open broadly with their tips extending widely. In the Thrombodes Complex of species, the flowers are most like those of *Enc. amicta.* Like the other species in this group, they show the dark maroon anther cap, the hairy base or callus of the lip, and the raised, even keeled, verrucose veins of the lip midlobe. They were described by Schlechter as being "semiauriculate" and as having smooth ovaries, not asperous. Schlechter did not mention any pubescence or hairs on the lower portions of the lip or callus, although they surely were there. These qualities show for the most part in his drawing #543 from Mansfeld's *Atlas der Kordillerenstaaten* (1929). A living plant bloomed at Dahlem Botanical Garden, Berlin, in June, 1914. It must have been the subject for the drawing #543 and provides us with an iconotype for the species. I find a difference in flower size among the clones I have seen.

The measurements Schlechter provides are leaves 25–35 cm long × 1 cm wide; sepals and petals 14 mm long; lip 12 mm long and 10 mm between tips of lateral lobes when the lip is flattened; midlobe about 8 mm long × 5 mm wide; and column 6 mm long. My notes from herbarium specimens indicate the sepals and petals are green, and the lip is white with dark maroon veining broken into verrucosities. The lip yellows with age and its rim is without colored veining. The lateral lobes are dark maroon veined over pale green, but have white edges at the tips. The flowers are produced on a long-stalked panicle at the height of the leaves. Little else about this species is known at the present. Schlechter notes that it is the southernmost species of *Encyclia* to be found.

Una Encyclia

Brazil

(FIGURE 1-7; PLATE 68)

Encyclia unaensis Fowlie. 1991. *Orchid Digest* 55:26.

This species comes from a unique area in coastal Brazil in Bahia state. A small nature preserve there is home for an unusual miniature marmoset, and the plant life is equally different. The area is near the home of *Laelia alaorii,* which is found on top of the Serra da Onca, and near the home of *Cattleya kerrii,* which grows along creeks from the coastal plain that empty into the Atlantic Ocean. It is an area less than 64 km long and wide, and it receives more rain than usual. It is south of the city of Ilheus, the cocoa center for the Western Hemisphere. The Serra da Onca also supports several Brazilian orchids that are otherwise found only in the wet Amazon regions. Not unexpectedly, this new species of *Encyclia* also was found there. The type specimen is Fowlie 90B5, collected along with Eclesio Hollanda da Cavalcante, Denis Duveen, and Earl Ross, and is preserved in the UCLA Herbarium. The plants were flowering in November, the Brazilian spring.

Fowlie's descriptive details for the plant are as follows: spherical to pyriform pseudobulbs to 2.6 cm tall, one to two leaves to 24 cm long × 1.2 cm wide with pronounced midrib grooves and acuminate tips; dorsal sepal oblanceolate, 15 × 4 mm, with subacute to rounded apex; petals oblanceolate, acute, narrowed in lower third to a claw, 12 × 3.5 mm; lip with lateral lobes straplike with rounded tips and reflexed backward to expose the column tip; midlobe ovate to rounded with apiculate tip, 6 mm in diameter with undulate margins. The sepals and petals are olive brown with a distinct yellowish border. The lip is a creamy white with seven bright magenta veins forming a dense central spot, and it has a wedge-shaped base. The column has a white apex and auricles. The 7–10 flowers are in a panicle, are not fragrant, and have a spread of 2.8 cm.

Yellow and Green Encyclia

Brazil

(FIGURES 1-7, 2-38; PLATE 69)

Encyclia viridiflava Menezes. 1991. *Orchid Digest* 55(1):21.

One must be careful not to confuse the *viridiflava* epithet of this species with the *viridiflora* of the following species. They are quite different. Lou Menezes describes this species as "spectacularly lime-green colored," and I, for one, look forward to seeing fresh flowers in the future. The plants come from Rondônia, a far western state of Brazil that is adjacent to Bolivia. The species is one of several discovered by Menezes with Jack Fowlie and Denis Duveen. Most of these *Encyclia* species have been published in the *Orchid Digest*, and a summary article has been published (Fowlie 1992). A few species have been published since 1992, mostly by Menezes, after the untimely deaths of the other two investigators.

Menezes and her colleagues found the new *Encyclia* species in the municipalities of Rolim de Moura and Ariquemes during the winter dry season (June, July, and August). The plants grew epiphytically, with flowering in June. The sepals and petals are firmly stiff and fleshy. The lip is 11 mm long with a spread of 8 mm across the lateral lobes. The midlobe is 6 mm wide and 5 mm long. Rick Cirino in Costa Mesa, California, has provided some fresh flowers, and I am at the last minute able to add a few more details and measurements to the description.

The plants are small with pseudobulbs 2.0–2.5 cm long and one or two upright leaves measuring 22 cm long × 1.8 cm wide. The flowers are in a simple, erect, fractiflex raceme somewhat shorter than the leaves and produce only five to seven flowers without fragrance. Vigorous plants have some branches and a few more flowers. The lime green sepals are oblanceolate, pointed, and 10 × 4 mm. The petals are narrower, about 3 mm. The lip lateral lobes have an oblong shape with rounded apices and are purple-veined at their bases on green-yellow backgrounds. The rounded and apiculate midlobe has a distinct isthmus with the two main callus ridges and the central vein extending as a rib to the apex. The three main veins are marked with red-purple on the disc and two more veins, one on either side, are faintly marked. The photographs show the column is yellow-green shading to green at its base, and the anther cap is yellow. The column is only 5 mm long and has definite auricles on the widened apex. The surface of the ovaries was irregularly roughened, and the flower stalks are smooth and green.

The orchid had also been found by Antonio Schmidt, owner of Chacara Bela Vista, a commercial orchid nursery in Assis, São Paulo, in a different part of Rondônia. He had found it near Nova Alvarada City. Castro (1998) considers this species synonymous with *Encyclia chlorantha* (please see). It does have similarities with this species.

Green-flowered Encyclia

Brazil

(FIGURES 1-7, 2-39; PLATE 70)

Encyclia viridiflora Hooker 1828. *Curtis's Bot. Mag.* t. 831.

This controversial species was the first *Encyclia* species to be collected and was named in 1828 by Hooker. It has never been collected a second time since William Harrison found it near Rio de Janeiro "wild in Brazil." Is it truly extinct, eliminated by the expansion of that exciting and beautiful city, or does it still reside in some small pocket or on some unscalable sugar-loaf rock? The type specimen is preserved in the Lindley Herbarium at Kew Gardens, and the illustration and description in *Curtis's Botanical Magazine* show its characteristics in good detail. But, we would like a second specimen to study after a gap of 170 plus years. In 1975 I obtained permission from Peter Taylor, the Head Keeper of the Orchid Herbarium then, and, with considerable trepidation, boiled up a single flower from the type specimen and prepared a drawing. That drawing (see Figure 2-39) was used in my description of the species in 1980, published in the *American Orchid Society Bulletin*.

Figure 2-38. *Encyclia viridiflava.* Drawing of the type specimen published in the *Orchid Digest,* 1991. With permission of the publishers.

As explained in Volume IV of this series, the flowers of *Encyclia viridiflora* are not the best models for a large genus such as *Encyclia.* Accordingly the species has been placed in a subgenus by itself. Among the differences that are quickly noted are the non-resupinate flowers, the shortness of the column, and the degree of lobing of the lip. Thomas Moore said it was an unattractive species with green flowers stained at the base with purple.

The deep purple anther cap had an unusual bilobed appearance, and the lip showed unusual lateral lobing that scarcely separated the midlobe. The callus was two-ridged or keeled, but it is not in the usual forcipate structure typical of so many encyclias. The pseudobulbs were described as being the size of nutmegs, and the plant had two leaves up to 17 cm long × 1.3 cm wide. Actual measurements of the flower parts show the dorsal sepal 9 × 2 mm, the petals narrower and curved, 8 × 1 mm, the lip 6 mm long × 4 mm wide, and the column 3–4 mm long. The flowers are among the smallest in the genus.

Where is that second plant to be found? Without it, we have a genuine extinct orchid for the records.

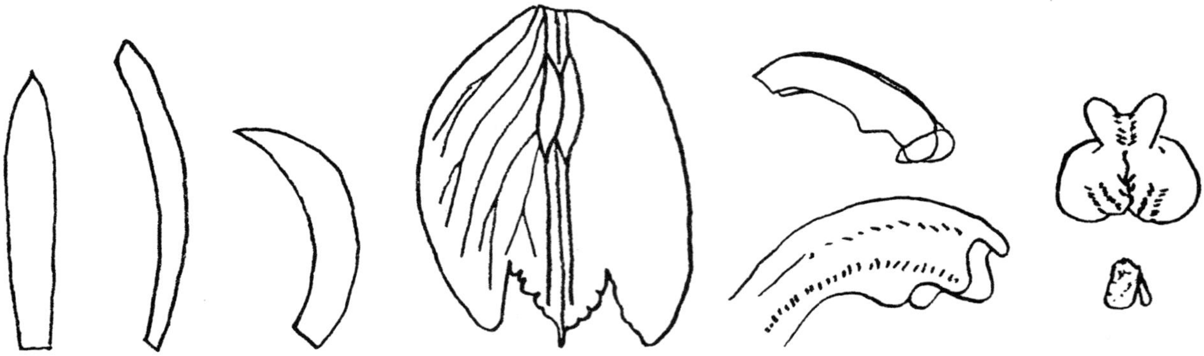

Figure 2-39. *Encyclia viridiflora.* Flower dissection and drawing by the author to show details of the type specimen, conserved in the Lindley Herbarium at Kew Gardens.

Desert Encyclia

Brazil

(FIGURES 1-7, 2-40; PLATE 71)
Encyclia xerophytica Pabst. 1976. *Bradea* 2(14):81.

Found on rocks in a xerophytic or desertlike environment in Bahia state, this species is well named. It was collected by Severiano and flowered in the collection of A. B. Gomez Ferreira in January, 1972. Its specific location was not revealed to prevent collectors from devastating the area there. The pseudobulbs were narrowly conical with three leaves that were linear-lanceolate, straight and fleshy, and 6–10 cm long × 1.0–1.5 cm wide. The 10-flowered raceme was 50 cm tall, the sepals and petals were olive green, and the lip was white with magenta veining.

The sepals were obtusely lanceolate, 25 × 7 mm, and the petals were linear-elliptic with narrow bases, about the same size. The lateral lobes of the lip were narrowly ovate, descending obliquely when flattened, and the midlobe was rounded, 12 mm long × 11 mm wide. The distance between lateral lobes was 20 mm. The apex of the callus extended into three veins with the central largest. The column was 12 mm with distinct auricles, and the pedicellate ovary was distinctly roughened, 23 mm long.

Without further information or plants to study we cannot know more.

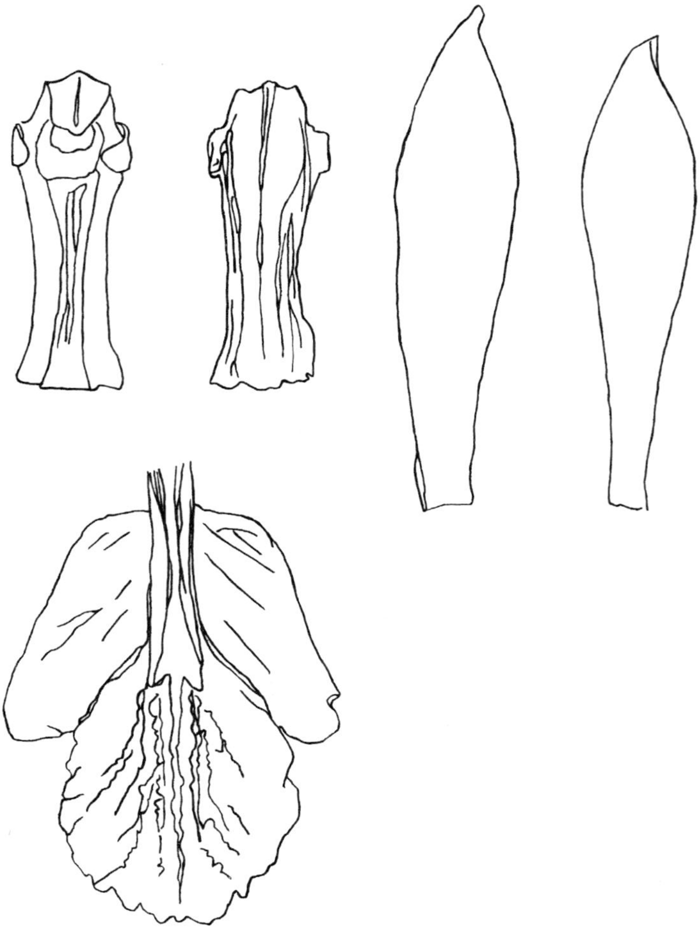

Figure 2-40. *Encyclia xerophytica.* A flower of the type specimen conserved in the Pabst Herbarium in Brazil. *Top row, left to right:* column from above, dorsal sepal, and petal. *Bottom row:* lip. Camera lucida drawing by Antonio Toscano de Brito.

Yauapery Encyclia

Bolivia, Brazil

(FIGURES 1-7, 2-41; PLATE 72)

Encyclia yauaperyensis (B. Rodrigues) Porto and Brade. 1935. *Rodriguésia* 1(2):29.

Synonym

Epidendrum yauaperyense B. Rodrigues. 1891. *Vellozia*, ed. 2, 1:122.

This species is named after Yauapery province in Alto Amazonas, Brazil. The name has been spelled jauperiensis, for instance, or Yaupery,

Figure 2-41. *Encyclia yauaperyensis.* Type drawing by Barbosa Rodrigues published as *Epidendrum yauaperyense* by Cogniaux. Adapted from Plate 10 of *Martius* 3(5).

but it is published as written by Rodrigues. It is a unique species, the only one so far in South America with a midlobe that is triangular, apiculate, and coming to a rounded or obtuse point. In this regard it is somewhat like *Encyclia tuerckheimii* from Mesoamerica (see Vol. V), also a species with that unusual midlobe shape, though narrower and more pointed. Fortunately for us *Enc. yauaperyensis* is pictured by Cogniaux in *Martius* (as Plate 10, fig. I) from Rodrigues's illustration.

The herbarium record has little to go on. According to Rodrigues's painting, the sepals and petals are olive green with shadings of yellow along the edges and at their bases. The midlobe is white with some reddish streaking and a yellow margin. The callus is also white with, some say, a spot of red on the apex of each of the callus ridges. The lateral lobes are creamy with red veins, as is the column. The red spots on the callus ridges do not show in the black-and-white figures of the *Martius* illustration, though they are present in the original Rodrigues painting and represent, so far as I can tell, the origins of the red veinings on the disc. They are not mentioned in the type description, so must not be particularly unique. Some living plants or new collections would now be in order. I have never seen the species, nor had Cogniaux when he described it for *Martius.*

The sepals and petals are about 18–20 mm long, oblong-lanceolate, the petals a little wider than the sepals. The lip is about 15 mm long. The lateral lobes are triangular-oblong and obtuse, and the midlobe larger and ovate-lanceolate, the tip forming a sharp point. The column is 6–7 mm long with auricles. The ovary surface is smooth. Confirming information and more about this species would be most desirable. It appears to be a rare item.

CHAPTER 3

Dubious Species of *Encyclia* in South America

Several new species of *Epidendrum* and *Encyclia* have been published in *Boletim Coordenadoria Associações Orquidófilas do Brasil* (*Bol. C.A.O.B.*) or *Orquidário,* Brazilian orchid publications. These are incomplete descriptions of new species and are therefore invalid. The epithets should not be used as they lack approved botanical status by the International Rules of Nomenclature. In particular, they may lack the Latin diagnosis that is necessary for the new description; or, they lack the citation of a type specimen and its location in a recognized herbarium. Orchidists are cautioned not to use names that lack valid status. That is not to say that the species so described by their author do not exist. It means that they do not yet have proper scientific names by which they may be identified.

Species names in *Encyclia* that are in this category and have come to my attention are as follows: *Enc. alta, Enc. aryana, Enc. casabranquensis, Enc. cecropiophila, Enc. cerradiphyta, Enc. claudiana, Enc. euglossa, Enc. feijoana, Enc. kundergraberi, Enc. lanceolata, Enc. limae, Enc. majellanica, Enc. odorifera, Enc. oliveirana, Enc. peresiana, Enc. relicta, Enc. rivularia,* and *Enc. tortilla.* Their relationship to the presently described Brazilian species is not known. It does, however, point out that we may have a long way to go in describing the rich array of Brazilian species in this genus.

The following species are confusing in their identity for a variety of reasons. Usually it is the loss or lack of a type specimen with which they may be compared, or the type description is insufficient to make a positive identification possible. Often there is no published or unpublished drawing of the specimen by its author or anyone else. A comparative lack of record prevents working out a proper identification. Sometimes certain tentative identifications are possible—as can be read in the following descriptions.

List of Dubious Species of *Encyclia* in South America

Herment's Encyclia

Guianas

Encyclia hermentiana Brongniart *ex* Joseph Neumann. 1845–1846. *Rev. Hort.,* ser. 2a, 4:158.

Four new species of encyclias were described in the article cited above of 1845–1846, *Encyclia hermentiana* being one. The four were plants found in the orchid collection of Mrs. Quesnel. No type description of these species exists, and so far as I can tell they were never validly published or illustrated, nor were type specimens prepared. Nevertheless, the names may appear in the literature from time to time. Brongniart states that the flowers of *Enc. hermentiana* are near to those of *Enc. chloroleuca* but larger and little different. Most today consider *Enc. hermentiana* to be the same as *Enc. chloroleuca* though there is no way to check on it further. I consider it a dubious species as exact identification is impossible.

Huebner's Encyclia

Brazil

Encyclia huebneri Schlechter. 1925. *Beih. Bot. Centralbl.* 42(2):105.

This species was collected in Amazonas state of Brazil. It had green-brown sepals and petals, a white lip, an auriculate column, and an obovate lip midlobe with an emarginate apex and verrucose thickened veins. The type plant was Huebner #69 destroyed during the Berlin bombing. No one has ventured a guess as to its present identity, so it must be placed with the dubious species. At least it was properly published. The type description says it came from Campos Salles-Campina and was flowering in September. We need to revisit the type locality, find other specimens, and then see where this orchid fits into the scheme of things. Schlechter, in his discussion of the species, says that *Encyclia huebneri* is similar to *Enc. diota* or, even better, to *Enc. trachycarpa.* These latter are both Mesoamerican species and appear today to our eyes as two distinctly different species. It is a little difficult to imagine now what Schlechter was seeing.

Insidious Encyclia

Colombia?

Encyclia insidiosa (Reichenbach f.) Schlechter. 1920. *Repert. Spec. Nov. Regni Veg.*, Beihefte 7:254.

Synonym

Epidendrum insidiosum Reichenbach f. 1877. *Linnaea* 41:80.

Reichenbach likened this species to *Encyclia replicata.* It was growing in the Bull collection and had been painted by John Day and then presumably sent to Reichenbach for identification. I have not discovered it among the Reichenbach specimens, though another has perhaps been more lucky. The type description tells us that the sepals and petals are olive brown and the lip is yellow veined in dark purple. The column is winged; the lip broadly ovate and apiculate; the sepals cuneate-oblong and pointed; the petals narrower and not spatulate; the lateral lobes linear, straplike, and the apices lobulate; the midlobe with radiating carinate veins and acutely pointed.

According to herbarium notes that I made in Vienna, the Reichenbach specimen #5181 is a sprig sent from the Bull collection, and #6798 is a single flower with no source indicated. They may represent what Reichenbach saw then. I considered both of them examples of *Encyclia diota*. In the same folder was a specimen labeled #13 from México, January, 1880, in the Backhouse collection; it was *Enc. atrorubens.* The *atrorubens* has sometimes been considered a variety of *Enc. diota* by some orchidologists, but in Volume V of this series the two taxa are considered separate species. In any case the problem of *Enc. insidiosa* is more or less resolved. It is probably not a South American species and should be added as a synonym under *Enc. diota* in Volume V.

Lax Encyclia

Encyclia laxa Schlechter. 1918. *Not. Bot. Gard. Dahlem-Berlin* 7:277. Also see Mansfeld, 1929, *Atlas der Kordillerenstaaten,* t. 46, fig. 184.

Misapplications (*auctor non*)

Epidendrum laxum Swartz. 1788. *Prodr.*, p. 125 = *Pleurothallis laxa* (Swartz) Lindley.

Epidendrum laxum Poeppig and Endlicher. 1838. *Nov. Spec. Plant.* 2:2 = *Epidendrum compressum.*

Epidendrum laxum var. *mocoanum* Schlechter. 1924. *Repert. Spec. Nov. Regni Veg.*, Beihefte 27:67 = ?*Encyclia candollei.*

This plant, named by Schlechter for its open arching panicle, flowered at Dahlem in April, 1918, but its origin was unknown. Schlechter says it was possibly from Brazil. His drawing of the floral details shows that this species is one of the few with "incomplete" lobing—if it may be called that. Other species with this type of lobing, discussed in Volume V, are *Encyclia spatella* and *Enc. candollei* / *Enc. flabellata.* This characteristic of incomplete lobing has not yet turned up in any known South American species. So far as I know, there is only the type record and the outline drawing of the *Enc. laxa* flower. The type specimen was destroyed in Berlin.

Schlechter suggested that it is close to *Enc. angustifolia,* which Swartz had described from Jamaica as *Epidendrum angustifolium,* or the Cuban *Enc. oxypetalum* (Lindley) Schlechter, but it is neither of these (see Vol. IV).

It had medium-sized flowers that were olive brown marked with yellow. The measurements and shapes described were as follows: sepals straplike, pointed, 16 mm long; petals oblong-spatulate and blunt, 14 mm long; lip 14 mm long and 13 mm wide above the middle; lateral lobes oblique, oblong, and obtuse; midlobe suborbicular; column 7 mm without auricles; and ovary surface smooth, 25 mm long.

I would say this specimen (species), based on the callus position, lobing, and the column without auricles, is more likely an example of *Encyclia candollei* from México (Vol. V). Unless modern examples of a plant with incomplete lobing of the lip and no auricles or wings on the column turn up from Brazil, we can probably cross this species off the South American list. A Peruvian plant named *Epidendrum laxum* by Poeppig and Endlicher in 1838 is not the same species, nor is its variety *mocoanum* by Schlechter, as is cited above.

Pointed-leaf Encyclia

Brazil

Encyclia oxyphylla Schlechter. 1925. *Beih. Bot. Centralbl.* 42(2):106.

It is doubtful that we can ever know what this species really was and whether it was truly an *Encyclia* species. It was collected in Amazonas as Huebner #79, was flowering in January, had flowers with green sepals and petals, and a white lip with brown side lobes, was striped with lilac, and had "2 Blättchen über der Säule"—whatever that may mean—petals, sepals, bracts? Schlechter further writes that he does not really know what to do with this plant as it is seemingly different from other encyclias, especially since the flowers are shorter than the leaves. I wonder if it was not some species of *Maxillaria* with a short inflorescence and if the two small leaves were not floral bracts of some sort. The type specimen was destroyed in Berlin, so no resolution of the question seems possible. Likely, it was not even an *Encyclia* species. The epithet has also been spelled with an *i* instead of a *y*.

Xipheres-like Encyclia

Brazil

Encyclia xipheroides (Kraenzlin) Porto and Brade. 1935. *Rodriguésia* 1(2):29.

Synonym

Epidendrum xipheroides Kraenzlin. 1896. *Gard. Chron.*, n.s., 20:63.

All that is known about this species is the type description. The type specimen was destroyed in Berlin. I have found no other records on it, nor any other specimens in the various herbaria I have visited.

Kraenzlin states in his account that the flowers appear like those of *Epidendrum xipheres* as pictured in Plate 111 of *Refugium Botanicum,* but his new plant had larger "bulbs and smaller flowers of a different color." The plants had narrow, thick leathery leaves and peculiar gray warts on the ovary. The sepals were lanceolate, pointed, and the petals linear, both colored dark green with dull purple veins. The lip was golden-colored with purple veins; the callus and column were white. Five major veins ran toward the apex of the lip. The side lobes of the lip were small and inconspicuous. The plant came from eastern Brazil and was flowering at Sander and Company in St. Albans.

The pseudobulbs measured 7 cm tall × 3.5 cm wide and bore two pointed leaves, stiff, tough, and channeled, 20 cm long × 1 cm wide. The raceme was thin and carried only a few flowers. The midlobe of the lip was broadly oblong, almost quadrate, the callus pronounced, extending into the five fleshy veins, and the column white and about 1.5 cm long. The plant was flowering in June. Auricles are not mentioned, a necessary detail, nor whether the ovary surface was rough or smooth. One can assume it was rough with its peculiar gray warts.

This type description is fairly thorough, as such things go, and it may well be possible to connect this species with one that was described earlier —before we are finished studying the Brazilian species—in which case it may become a synonym. Another distinct possibility is that the origin of the plant was mistaken—Kraenzlin merely wrote, "said to be of Brazilian origin." The description of the plant and flowers, also the unique warts on the ovary, would match *Encyclia nematocaulon* in Mesoamerica and Cuba very well, and that would straighten out this problem of a dubious species. I will opt for this resolution of the identity of this species for the time being (see *Enc. nematocaulon* in Vol. V). We have nothing better to guide us.

CHAPTER 4

Biogeography and Evolutionary Relationships of the Laeliinae

A paper written by Lyman Smith (1962) on the origin of the flora of southern Brazil, which I read and studied many years ago, stimulated me to write down what I knew of the biogeography of the Laeliinae at that time. An earlier attempt (Withner and Adams, 1960) had resulted in a closer look at generic relationships in the subtribe, and it is appropriate to bring the material to date as a chapter for this set of books. Although some of the research has been known to orchidologists for many years (Brieger 1960a, 1960b, 1961, 1973), it is still relatively unknown, especially to others interested in plant geography. Biogeography is a currently popular subject as the result of the study of plate tectonics (Merritt, 1979; Scotese, 1990) and of having a greater understanding of the forces affecting plant speciation and evolution. Kirschvink and Ripperdam (1997) feel that the earth's polar continents moved into new positions, distributing weight to equatorial areas, and thereby created new environments. At that time, approximately 500 million years ago, there was an unusual burst of species formation, they estimate 20 times the usual, that correlates with the new positioning of the continents. The environmental upheavals that occurred included pronounced temperature changes and water level alterations.

The major specific publication for botanists in this field has been by Raven and Axelrod (1974): "Angiosperm Biogeography and Past Continental Movements." It was followed by another paper in 1975. These two authors summarize what was known at that time about patterns of plant distribution that emerge from various geological and biogeographical studies over the whole world and cover the various time periods and epochs. Orchids follow the same "rules" and migration pathways as other plants, and by applying the information to the various members of the Laeliinae we may learn more about their evolutionary relationships and the significance of their current distribution patterns. Unfortunately, the existence of orchid fossils is questionable or nonexistent so nothing but modern comparative anatomical, taxonomic, or biochemical studies are available to provide evolutionary evidence.

Pertinent literature includes the work of Garay (1960, 1963), who published much of the early data on the origin of the Orchidaceae as a complex of subfamilies, and who correlated certain distribution patterns with orchid evolution. Most recently Benzing (1987) has reviewed the various processes and patterns at work in orchid evolution—from geography and euglossine bees to the epiphytic habit—focusing more intently on some of the material covered by Dressler (1981) in his book on orchid natural history and classification. A newer book by Dressler (1993) also provides further data through the use of computers and cladistic analysis.

Cladistics is a newer type of classification system strictly based on evolutionary descent from a common ancestor. In addition to the more usual older Linnean systems of classification, cladistics can also be employed to correlate the evolution of a specific group of orchids with its geographic distribution. Cladistics produces schemes that depend upon the occurrence of shared characteristics derived from the common ancestry. It may also, by its strict objectivity, produce different or new views on the patterns of evolution within the orchid or other plant families. To understand some of these technicalities, the reader is referred to the work of Burns-Balogh and Funk (1986) or the article by Cracraft (1983) about attempts to reconstruct evolutionary systematic patterns in any plant taxon. Nevertheless, we will see more about such family trees, called cladograms, as research in cladistics continues and is applied to evolutionary problems in orchid speciation. Many have heard to date about the studies on the genus *Oncidium* (Chase and Hills, 1992), and currently the work is being extended to other genera (Pridgeon and Chase, 1998; Higgins, 1999).

Cladistics does not deal readily with hybridization as a source of variation or speciation, and this may cause some difficulties with the use of the technique within the orchid family. Orchids hybridize in nature, and there is no evidence they did not hybridize in the past, especially in the Laeliinae. Hybridization must be considered in cladistic research as a factor in orchid evolution, particularly in groups of closely related species or even genera.

Major orchid family distribution patterns emerge from all the information sources mentioned above. Most are dependent upon the ancient geological configurations and connections that were once existing among our present-day continents. The tropical areas were much more widespread than they are today. See the generalized maps of Pangea (Fig. 4-1) and the formation of Laurasia and Gondwanaland (Fig. 4-2) as rift systems began to split the original land mass of the earth into migrating continental and suboceanic plates. Scotese's (1990) work on the paleomapping project provides current detail.

Garay, following Takhtajan's ideas, points out that earliest migrations of orchid plants were from east to west, and they first involved terrestrial orchids, species with the most primitive of the various sympodial orchid growth habits. Raven and Axelrod (1974) believe the coordinated botanical and geological evidence for all flowering plants points to western Gondwanaland as the site of origin for ancestral angiosperms, including terrestrial orchids. Garay, however, indicates a Malaysian–Southeastern

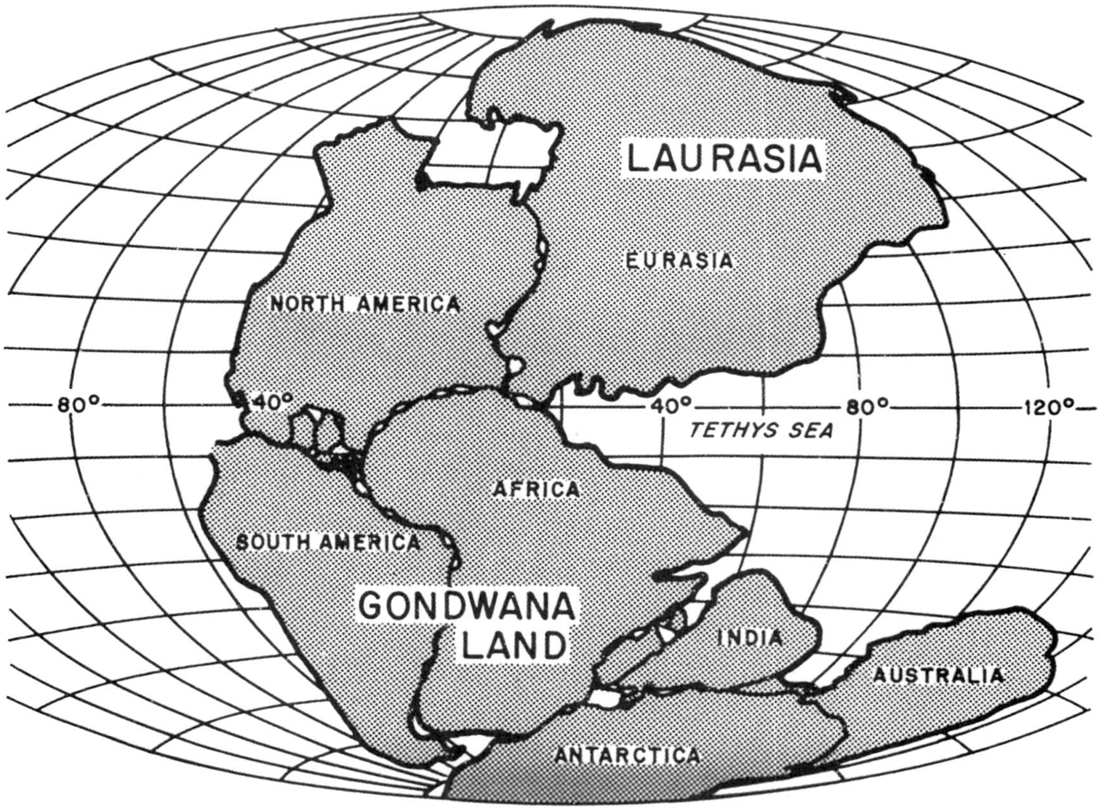

Figure 4-1. Reconstruction of the world's land area, Pangea, at the beginning of the Mesozoic Era, 240 million years ago. After Hurley (1968) and others. Drawing by Joy Dabney.

Asian point of origin for the orchid family. In either case, as plants spread and as continents drifted to new locations, the orchids also evolved. New species formed and ancestral forms died out. The epiphytic habit developed, and monopodial growth patterns emerged. New secondary centers of speciation developed from which radiation and migration of species to new locations took place.

The evolution of the early orchids must have coincided with the evolutionary development of certain groups of pollinating insects. With continued adaptive modifications of the plant body and the flowers, the ancestral orchid lines became even more specialized and occupied still other new environments. The orchids and other monocots capitalized mostly on the sympodial growth pattern (Holtum, 1955) with the formation of water-storing tissues in the stems and leaves. The development of the epiphytic habit, correlated with the evolution of insects and angiosperm shrubs and trees, produced additional distribution patterns and more diverse and specialized vegetative and floral variations. In fact, the term "explosive," attributable to Brieger (1960a), has been used many times to

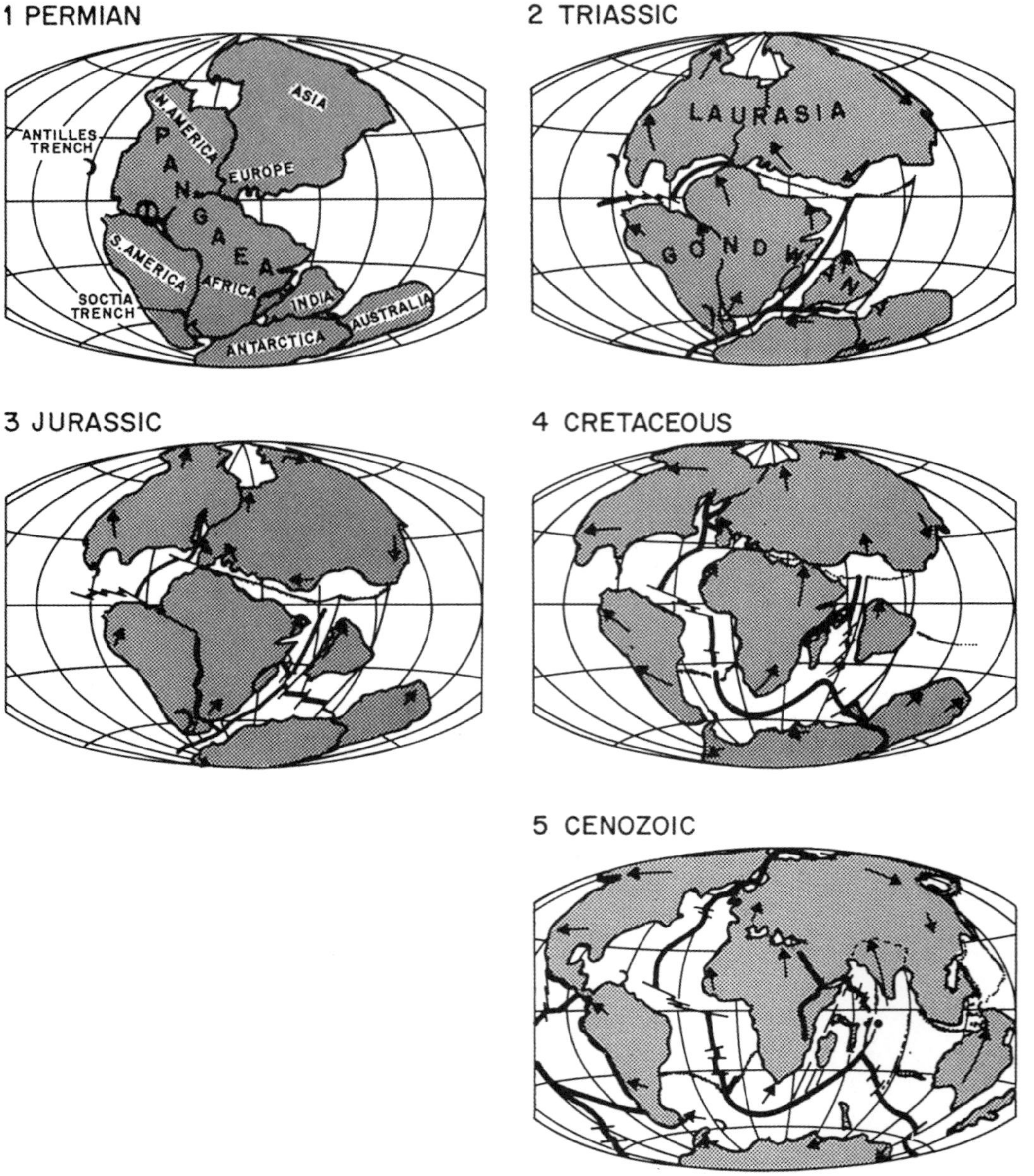

Figure 4-2. Formation and development of the rift systems as they begin to separate the continents. Arrows show the direction of movements; heavy solid lines the rifts. After Ben-Avraham (1981) and others. Drawing by Joy Dabney.

describe the proliferation and radiation of modern orchid genera and species in the last few thousand years, which geologists refer to as the Pleistocene Epoch.

As orchid evolution proceeded, certain geological and climatological alterations also took place. Since we concern ourselves here with the development of the various genera of the Laeliinae, this discussion is limited to changes in the New World. We cannot be certain how or when the ancestors of the Laeliinae migrated into South America. Three alternatives provide possible pathways, either directly or in combination. A major choice is through the early direct connections between Africa and North and South America when they were still part of Gondwanaland. Second, a temporary, later pathway also existed through Antarctic connections from what is now Malaysia, or what was eastern Laurasia in ancient times, or even earlier in western Gondwanaland. A third pathway is provided by long distance dispersal over water, though it is not as favored as the other two possibilities. Garay (1960) pointed out that only 32 of the approximately 800 orchid genera were involved in some kind of transoceanic distribution. These 32 genera represent 276 terrestrial orchids, indicating a migration that took place early in the history of the orchid family.

Once in the new South American locations, however, the ancestral plants eventually spread north through what became the Andean Cordilleran plates. Orchid ancestors were found in mountain habitats, adapting to various environments and profound climatic changes, diversifying and radiating into new areas and into a host of new genera and species. Most species tended to stay in higher areas, but a few adapted to the lowland distribution pattern. Concentrations of orchid species tend to follow the 1000 m contour in México, a finding that probably applies to the rest of the New World. It emphasizes their mountain origins. Though some species ultimately adapted to lowland areas, the Laeliinae mostly remained in montane habitats. Whatever their ancestors may have been, possibly some *Dendrobium*-like forms, there are no living present-day remnant taxa in Asia, Africa, or elsewhere to connect with this group. Today, the Laeliinae are only to be found in the neotropics of the New World.

PATHS OF MIGRATION

Spreading to the north after establishing Andean habitats, as indeed other plants such as *Araucaria angustifolia* (Parana pine), *Nothofagus* (southern beech) species, or *Gunnera* species have done (Raven and Axelrod, 1974), certain ancestral Laeliinae must have followed a line of migration that extended from southern Peru to the southern states of Brazil (Fig. 4-3). In southern Brazil an additional diversification occurred, the species extending in a northern or southern direction in the Brazilian coastal rain forests and the mountains that paralleled the coastal areas. Smith (1962) pointed out how an eventual break between Peru and Brazil had been brought about by environmental factors of heat, extending southward from the Amazon Basin, and aridity, pushing northward from southern Argentina. These two environmental factors, the result of marked geological and climatological alteration over time, effectively disrupted any east-

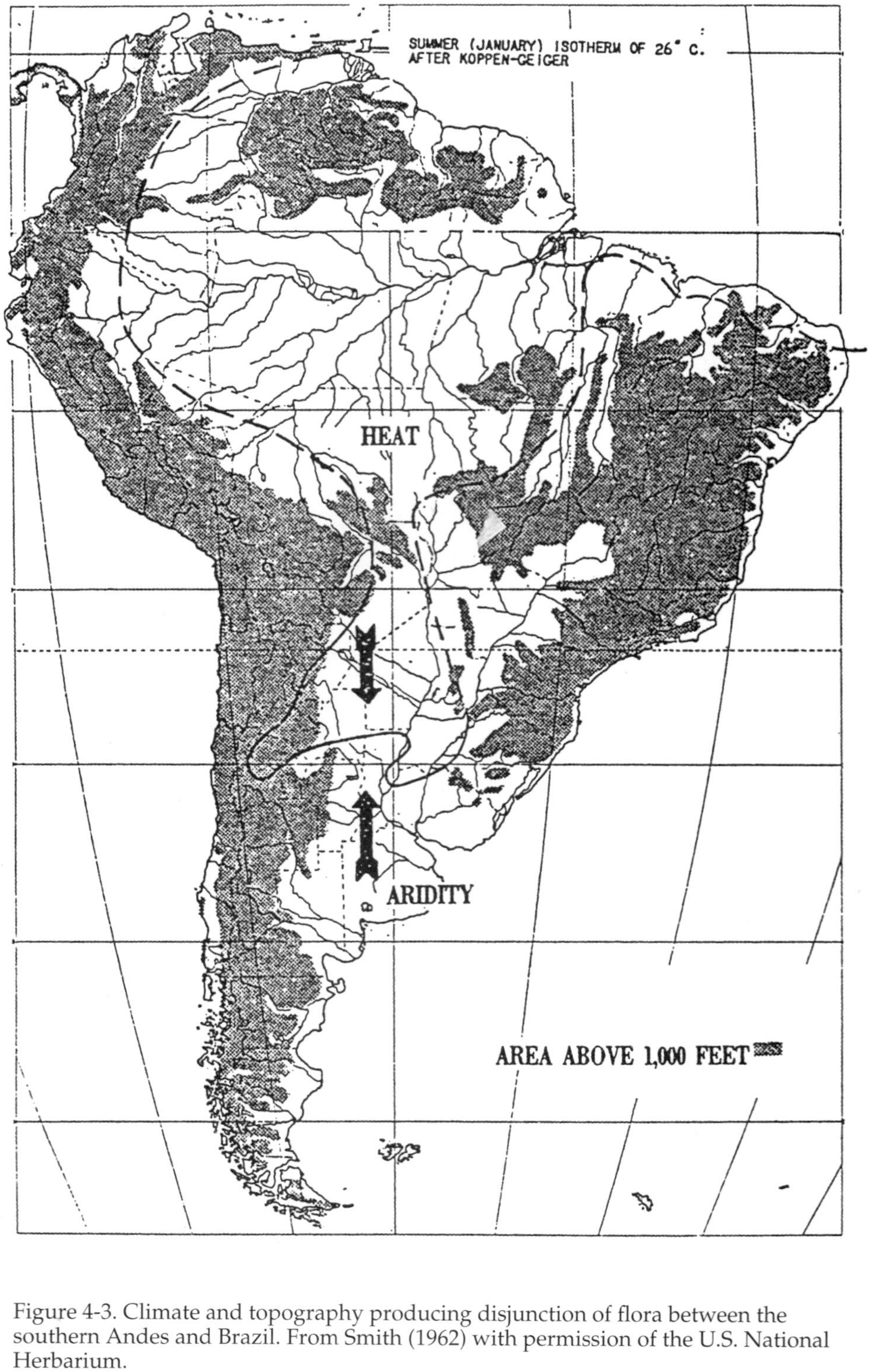

Figure 4-3. Climate and topography producing disjunction of flora between the southern Andes and Brazil. From Smith (1962) with permission of the U.S. National Herbarium.

west continuity in plant distribution by this pathway, especially for species originating in coastal Peru.

This kind of change can result in disjunct (separated) distribution patterns within a genus or for a species. The disjunction can then be followed by a new complex of genera or species that may form as the environment changes and an isolated population adapts to new conditions. In the Gran Chaco, as the hot, dry disruptive area, mostly in Bolivia, is now called, some of the highest temperatures reported in the New World have been recorded. The Chaco has been an outstanding climatological and geological feature for thousands of years. The Bolivian orchid flora is only now being studied intently (Bennett and Christenson, 1995), and it remains an exciting region for continued field studies and research.

Prance (1973, 1985) and others (see, for example, Haffer, 1969) have amassed evidence for another line of migration. They have noted that certain genera and species of various plants, butterflies, lizards, and birds have spread eastward from the more northern portion of the Andes to the northern coastal Brazilian mountains. Further diversification of species then took place, probably the same sort of processes that occurred in more southern Brazil. Fewer species migrated along the mountains of northern South America into the Guianas or migrated as far as what is now Trinidad and Tobago.

Mountain forests (Cretacean) provided the original habitat for these pathways from, or to, the Andes. All there is left today of these original wooded areas, after many tropical climatic and geological changes over the centuries, is a series of what are called refugia (Fig. 4-4). These are scattered pockets of primitive forest and animal species, or their descendants, that botanists and zoologists have discovered still existing despite general vegetation changes around them. Ice ages caused pronounced sea level variations, land level alterations occurred, and distinct climatic changes took place so that many new mountain formations and other geological transformations prevailed. The Andes rose mostly during the Middle Cretaceous, for instance.

The main line of migration indicated by the refugia more or less followed the southern edge of the present Amazon Basin and represents the northern edge of the ancient continental portion called the Brazilian Planalto. Some biota, however, migrated along the northern edge of the Basin and were thus distributed upon the Guianan Shield and around the Orinoco Basin. Numerous examples exist today of plants or animals limited in their distributions since the Middle Cretaceous to one side of the Amazon area or the other, or to one river system, and these correlate nicely with the early geological history of the region. Still other refugia are located in northern Colombia, relating to the termination of the central Andean range of mountains in an area that is still of uncertain geological history (Mooney, 1981).

The refugia populations of what were once more widespread plant species may undergo a variety of changes. Certain species may become extinct. Others may begin to differentiate from the original species by hybridization and segregation or selection until new species are formed. One is tempted to believe that orchids of past ages behaved in essentially the same way as they do in nature today, though there is no study on this

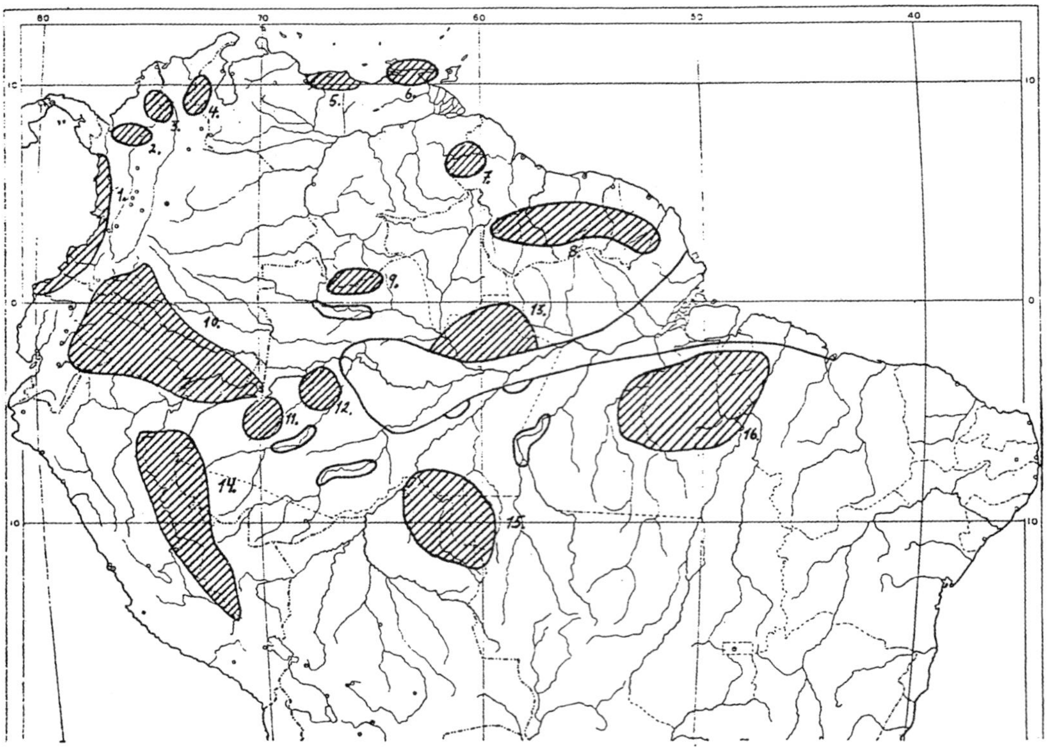

Figure 4-4. Forest refugia proposed by Prance (1985): 1, Chocó; 2, Nechi; 3, Santa Marta; 4, Calumbo; 5, Rancho Grande; 6, Paria; 7, Imataca; 8, Guiana; 9, Imerí; 10, Napo; 11, Olivença; 12, Teffé; 13, Manaus; 14, East Peru; 15, Rondônia-Aripuana; 16, Belém-Xingú. Permission of *Acta Amazonica,* Inst. Nac. de Pesquisas, Brazil. Used with permission.

point that is directly applicable. The rock laelias (Vol. II) especially fit into this mountain context and seem to be caught today in a period of transition, speciation, and expansion in their mountain habitats in eastern Brazil. One can also note that most small or monotypic genera of the Laeliinae are either found in central or southern Brazil or on the Caribbean Islands, a kind of disjunction. Most are also rather specialized in being miniature orchids. It is difficult to say whether they represent the relict remnants of once more widely distributed species that migrated the farthest—or whether they are concentrations of comparatively new genera as a result of the earlier isolation of their ancestors. I am inclined to the latter view, though I cannot back it with any direct evidence at this time. Probably both explanations can be applied.

Camp (1947) stated that the present Amazon flora is a heterogeneous group of plants derived from the surrounding upland areas. He also pointed out that the flora of the Orinoco and Amazon Basins shows its more recent character by having poorly delimited species that have resulted from "yet active genic introgressions and unstabilized segregations." He felt the Amazon flora, because of the region's relatively recent emergence in Middle Cretaceous times following the inundations of the Tertiary, did not go back much later than late Pliocene or late Pleistocene about 10–12 million years ago. With very few exceptions (perhaps 5–6 among more than 200 species), the epiphytic species of the Laeliinae are not found in these once submerged, lowland basin areas. This pattern, in turn, points to a still earlier origin and distribution for the subtribe. The ancestral orchids of the Laeliinae must have existed by the Early or at least the Middle Cretaceous to produce the current distribution patterns.

Yet another line of migration, after following the Andes northward from Peru, extended eastwards from present Colombia toward the coastal mountainous areas of Venezuela, then the Guianas, and terminated finally in Trinidad. Some of these mountains are submerged so that the connection to Trinidad no longer exists, but the orchid and other flora of that island show a direct connection once existed.

All these migratory pathways demonstrate the importance of the ancestral Andean flora as a source for South American Laeliinae as well as other epiphytic orchids. It is not surprising therefore that Peru, Ecuador, and Colombia are still rich treasure houses of many unique genera and species. I have noted elsewhere that the fieldwork and taxonomic research of Dodson (1988) and his colleagues in Ecuador have shown that small country, about the size of the state of Colorado in the United States, now has at least 3000 orchid species. That number of species is approximately 10 percent of the whole orchid family, as modern estimates indicate 30,000 species for the Orchidaceae! One assumes that these species-rich Andean areas not only gave rise to the migrant ancestry that crossed South America to Brazil or Trinidad, but they also had their own secondary diversification of species following the localized geological and climactic changes that took place after the South American continent was finally formed. Not all the ancestral orchids moved on to distant locations into eastern South America, and some plants came directly when Africa and South America were connected.

From South America the pathway of distribution of the Laeliinae extends from Colombia northwards through the mountains of Central America to México with various indications of migration eastward from Yucatan to the islands of the Caribbean and to Florida and the Bahamas. In México, as a northern extension, the mountain chain branches to form a large Y, one branch paralleling the Caribbean, the other the Pacific coast. There were no certain connections between South and Central America until the Early Miocene, about 20.5 million years ago. Finally, some northern terrestrial species migrated through México into South America. These movements did not, however, involve the Laeliinae, so that southward migration does not seem pertinent to this discussion. The ancestral Laeliinae seem to have primarily moved from south to north, and from west to east within the New World.

The processes of modern continent formation, as we know them today, began in the Middle Cretaceous about 100–110 million years ago (Fig. 4-2), and certain ancestral orchids of a variety of types must have already existed for modern orchids to be in present locations. This conclusion seems to be an inevitable one when the geological maps and orchid distributions are correlated. It is a topic of particular interest to geologist-botanist Clair Ossian (work in progress), and we look forward to his conclusions, especially since most botanists today claim that flowering plants evolved during the Middle Cretaceous. It is difficult to see how orchids, especially, could have already been on a variety of continents in the Middle Cretaceous when presumably more primitive plant families were just beginning to evolve—another area for further research.

We perhaps end up with more questions than have been resolved, but as knowledge of geology increases, we try to keep pace with our interpretations of plant distributions and evolution. We know that South America parted from Africa about 90 million years ago, after which direct migration and interchange of flora between the two continents became unlikely. We also know that tropical and subtropical species, in spite of water barriers, could have spread earlier to South America from India, Antarctica, and what became the Scotia Arc, with distinct floras already existing along the way. Later more temperate or cool-growing species spread from Australasia to South America about 40–45 million years ago. Raven and Axelrod (1974) also feel that evidence based on fossil pollen shows that modern angiosperm genera existed 65–70 million years ago, and that ancestral forms may have come as early as 110 million years ago. Since some genera had reached the West Indies by 40–45 million years ago, most migration of the Laeliinae must have occurred in the interval from 100 to 40 million years ago, an interval of 60 million years that spans the Upper Cretaceous and Lower Tertiary periods.

When more studies are carried out with research on chloroplast and nuclear DNA, amino acid sequences, photosynthetic patterns, gene structure, isozymes and other enzyme peculiarities, pigments in flowers and leaves, and particular compounds in fragrances, nectars, and plant cells, we will gradually be better able to point out past relationships. We will more closely come to reconstructing the family tree of the Laeliinae and the other tribes and subtribes, fitting them into the geological history that is already reasonably well known and dated.

APPENDIX 1

Taxonomic and Nomenclatural Changes in Volumes I–VI

VOLUME I. THE CATTLEYAS

p. 21 *Cattleya* subg. *Laelioidea* Withner. 1988. *Subg. nov.*
p. 21 *Cattleya* subg. *Rhizanthemum* (Cogniaux) Withner. 1988. *Subg. nov. Comb. nov.*
p. 21 *Cattleya* subg. *Cattleya* section *Xantheae* Withner. 1988. *Sect. nov.*
p. 21 *Cattleya* subg. *Cattleya* section *Maximae* Withner. 1988. *Sect. nov.*
p. 21 *Cattleya* subg. *Stellata* Withner. 1988. *Subg. nov.*
p. 21 *Cattleya* subg. *Circumvolva* Withner. 1988. *Subg. nov.*
p. 21 *Cattleya* subg. *Circumvolva* section *Aurantiacae* Withner. 1988. *Sect. nov.*
p. 21 *Cattleya* subg. *Aclandia* (Pfitzer) Withner. 1988. *Subg. nov. Comb. nov.*
p. 21 *Cattleya* subg. *Intermediae* (Cogniaux) Withner. 1988. *Subg. nov. Comb. nov.*
p. 21 *Cattleya* subg. *Bicolorea* (Pfitzer) Withner. 1988. *Subg. nov. Comb. nov.*
p. 21 *Cattleya* subg. *Leopoldia* (Pfitzer) Withner. 1988. *Subg. nov. Comb. nov.*
p. 21 *Cattleya* subg. *Leopoldia* section *Granulosae* Withner. 1988. *Sect. nov.*
p. 95 *Cattleya mooreana* Withner, Allison, and Guenard. 1988. *Sp. nov.*

VOLUME II. THE LAELIAS

p. 18 *Laelia* subg. *Crispae* (Pfitzer) Withner. 1990. *Comb. nov.*
p. 18 *Laelia* subg. *Crispae* section *Perriniae* Withner. 1990. *Sect. nov.*

p. 18 *Laelia* subg. *Crispae* section *Sinocoranae* Withner. 1990. *Sect. nov.*
p. 18 *Laelia* subg. *Microlaelia* (Schlechter) Withner. 1990. *Comb. nov.*
p. 18 *Laelia* subg. *Parviflorae* (Lindley) Withner. 1990. *Comb. nov.*
p. 18 *Laelia* subg. *Parviflorae* section *Harpophyllae* Withner. 1990. *Sect. nov.*
p. 18 *Laelia* subg. *Parviflorae* section *Esalqueanae* Withner. 1990. *Sect. nov.*
p. 18 *Laelia* subg. *Parviflorae* section *Rupestres* Withner. 1990. *Sect. nov.*
p. 18 *Laelia* subg. *Parviflorae* section *Liliputanae* Withner. 1990. *Sect. nov.*
p. 48 *Laelia brevicaulis* (H. G. Jones) Withner. 1990. *Comb. nov.*
p. 122 *Laelia sanguiloba* Withner. 1990. *Sp. nov.*
p. 127 *Laelia spectabilis* (Paxton) Withner. 1990. *Comb. nov.*

VOLUME III. *SCHOMBURGKIA, SOPHRONITIS,* AND OTHER SOUTH AMERICAN GENERA

p. 21 *Schomburgkia albopurpurea* (Strachan *ex* Fawcett) Withner. 1993. *Stat. nov.*
p. 41 *Schomburgkia sawyeri* (Williams) Withner. 1993. *Comb. nov.*
p. 77 *Sophronitis pygmaea* (Pabst) Withner. 1993. *Comb. nov. Stat. nov.*

VOLUME IV. THE BAHAMIAN AND CARIBBEAN SPECIES

p. 64 *Encyclia ochrantha* (Richard) Withner. 1996. *Comb. nov.*
p. 116 *Tetramicra* subg. *Tetramicra* section *Parviflora* Withner. 1996. *Sect. nov.*
p. 116 *Tetramicra* subg. *Tetramicra* section *Ekmanii* Withner. 1996. *Sect. nov.*
p. 116 *Tetramicra* subg. *Pachycaula* Withner. 1996. *Subg. nov.*

VOLUME V. *BRASSAVOLA, ENCYCLIA,* AND OTHER GENERA OF MÉXICO AND CENTRAL AMERICA

p. 104 *Encyclia cordigera* f. *leucantha* (Schlechter) Withner. 1998. *Comb. nov. Stat. nov.*
p. 121 *Encyclia parviflora* (Regel) Withner. 1998. *Comb. nov. Stat. nov.*
p. 123 *Encyclia perplexa* (Dressler and Pollard) Withner. 1998. *Stat. nov.*
p. 137 *Euchile* (Dressler and Pollard) Withner. 1998. *Genus nov. Stat. nov.*
p. 138 *Euchile citrina* (Dressler) Withner. 1998. *Comb. nov. Stat. nov.*
p. 139 *Euchile mariae* (Ames) Withner. 1998. *Comb. nov. Stat. nov.*
p. 152 *Homalopetalum* subg. *Pachyphyllum* Withner. 1998. *Subg. nov.*
p. 158 *Homalopetalum kienastii* (Reichenbach f.) Withner. 1998. *Comb. nov.*

VOLUME VI. THE SOUTH AMERICAN *ENCYCLIA* SPECIES

p. 75 *Encyclia garzonensis* Withner. 2000. *Sp. nov.*
p. 91 *Encyclia magdalenae* Withner. 2000. *Sp. nov.*
p. 119 *Encyclia quesneliana* (Hort. ex Henshall) Withner. 2000. *Comb. et stat. nov.*
p. 122 *Encyclia randiana* (B. Rodrigues *ex* Linden and Rodigas) Withner. 2000. *Comb. nov.*

APPENDIX 2

Additions and Corrections to Volumes I–VI

VOLUME I. THE CATTLEYAS

p. 5 The proper date for Lindley's publication of the genus *Cattleya* appears to be 1821, not 1824 nor 1825. The date of 1821 has not yet been confirmed by librarians that specialize in such matters, but it does seem likely.

p. 10 Leslie Garay reports that Pfitzer (1889) antedated Rolfe (1895) in attempting to classify the cattleyas. Accordingly, some of my terminology for subgenera and sections of *Cattleya* should be modified to make them accurate according to the rules of priority. Some changes involve spelling, others different epithets as they are used in Volume I. See changes below for pages 19 and 21. The Pfitzer reference is in the bibliography of Volume VI.

p. 14 Add the following intergeneric combinations with *Cattleya* to Table II, bringing the table up-to-date through November, 1999.

× *Barkeria* × *Brassavola* × *Laelia* × *Sophronitis* = *Trisuloara (Tsla.)*
× *Barkeria* × *Broughtonia* = *Turnbowara (Tbwa.)*
× *Barkeria* × *Laelia* × *Sophronitis* = *Matsudaara (Msda.)*
× *Brassavola* × *Broughtonia* = *Carmichaelera (Crml.)*
× *Brassavola* × *Broughtonia* × *Laelia* × *Schomburgkia* = *Westara (Wsta.)*
× *Brassavola* × *Broughtonia* × *Laelia* × *Schomburgkia* × *Sophronitis* = *Mooreara (Mora.)*
× *Brassavola* × *Diacrium* × *Sophronitis* = *Clarkeara (Clka.)*

× *Brassavola* × *Epidendrum* × *Laelia* × *Schomburgkia* × *Sophronitis* = *Johnyeeara (Jya.)*
× *Brassavola* × *Epidendrum* × *Schomburgkia* = *Symmonsara (Syma.)*
× *Broughtonia* × *Diacrium* × *Laelia* = *Cookara (Cook.)*
× *Broughtonia* × *Diacrium* = *Kraussara (Krsa.)*
× *Broughtonia* × *Laelia* × *Laeliopsis* = *Fialaara (Fia.)*
× *Broughtonia* × *Laeliopsis* × *Tetramicra* = *Fordyceara (Fdea.)*
× *Diacrium* × *Laelia* × *Schomburgkia* = *Mailamaiara (Mai.)*
× *Tetramicra* = *Tetracattleya (Ttet.)*

p. 19 Change *Rhizantha* to *Rhizanthemum.*
Add *silvana* to the list of species under Section *Cattleya.*
Switch order of *Aurantiacae* and *Moradae,* and change *Moradae* to *Circumvolva.*
Change *Intermedia* to *Intermediae.* Move *dolosa* to Section *Rhizanthemum.*
Change *Schomburgkoidea* to *Bicolorea.*
Change *Falcata* to *Leopoldia.*
Change *Guttatae* to *Leopoldia.*

p. 21 Changes to *Cattleya* classification scheme based on Pfitzer (1889):
Change *Rhizantha* to *Rhizanthemum*
Rhizanthemum (Cogniaux) Withner. *Subg. nov. Comb. nov.* Basionym: Subsection *Rhizanthemum* Cogniaux. 1898–1902. In *Martius* 3(5):189. Type: *C. walkeriana* Gardner.
Change *Moradae* to *Circumvolva* Withner and move the new section *Circumvolva* ahead of *Aurantiacae.*
Aclandia (Pfitzer) Withner. *Subg. nov. Comb. nov.* Basionym: Section II. *Aclandia* Pfitzer. 1889. In Engler and Prantl's *Die Naturliche Pflanzenfamilien. Orchidaceae* 2(6):146. Type: *C. aclandiae* Lindley.
Change *Intermedia* to *Intermediae.*
Intermediae (Cogniaux) Withner. *Subg. nov. Comb. nov.* Basionym: Section *Intermediae* Cogniaux. 1898–1902. In *Martius* 3(5):189. Type: *C. forbesii* Lindley.
Change *Schomburgkoidea* to *Bicolorea.*
Subgenus *Bicolorea* (Pfitzer) Withner. *Subg. nov. Comb. nov.* Basionym: Section I. *Bicolorea* Pfitzer. 1889. In Engler and Prantl's *Die Naturliche Pflanzenfamilien. Orchidaceae* 2(6):146. Type: *C. violacea* (Kunth) Rolfe.
Change *Falcata* to *Leopoldia.*
Subgenus *Leopoldia* (Pfitzer) Withner. *Subg. nov. Comb. nov.* Basionym: Section IV. *Leopoldia* Pfitzer. 1889. In Engler and Prantl's *Die Naturliche Pflanzenfamilien. Orchidaceae* 2(6):146. Type: *C. leopoldi* Verschaff.
Change section *Guttatae* to *Leopoldia* Withner.

p. 36 *Epidendrum aclandiae* (Bateman *ex* Lindley) Reichenbach f. 1861.

p. 37 *Epidendrum amethystoglossum* (Lindley and Reichenbach f.) Reichenbach f. 1861.

p. 40 *Cattleya aurantiaca* (Bateman *ex* Lindley) Don. 1840.

p. 40 Add under synonyms:
Broughtonia aurea Lindley. 1840. *Bot. Reg.* misc. 22.

Epidendrum aureum (Lindley) Lindley. 1853.

Laelia aurantiaca (Lindley) Beer. 1854. *Prakt. Stud. an der Fam. Orchidaceae*, p. 256.

p. 42 *Cattleya aurea* (T. Moore *ex* Warner and Williams) Rodigas. 1881.

Cattleya dowiana aurea T. Moore *ex* Warner and Williams. 1883.

Cattleya chrysotoxa (Sander) Godefroy-Leboeuf. 1892. *L'Orquidophile*, p. 31.

p. 44 *Epidendrum bicolor* (Lindley) Reichenbach f. 1861.

Add under synonyms:

C. brasiliensis Klinge. 1898. *Acta Hort. Petróp.*17:7, fig. 1.

C. bicolor var. *measursiana* Warner and Williams. 1889. *Orchid Album* 8, pl. 357.

C. measursiana (L. O. Williams) Blumenschein. 1961. *Publ. Cent. Inst. Genetics*, Piracicaba, Brazil 2:23–33. Note that Blumenschein here pointed out the tetraploid nature of the Minas Gerais populations of *C. bicolor*, saying that they corresponded to the var. *measursiana*. Accordingly, Blumenschein raised the variety to species status because of the different number of chromosomes, flower and plant size, and the difference in flowering time.

p. 46 *Cattleya* ×*joaquiniana* Miranda. 1999. *Bradea* 8 (22): 128–130. The type specimen is Miranda #1419 at Brasilia, flowering in May on Córrego do Soldado. This new natural hybrid between *C. walkeriana* and *C. bicolor* has flower and vegetative characteristics intermediate between the two parents. The lip is distinctly trilobed with an elongated isthmus. Both parents are abundant in its habitat in Minas Gerais state, Brazil, in gallery forests at 900 m. See figure p. 170.

p. 46 *Cattleya autumnalis* Hort. *ex* O'Brien. 1855.

p. 47 *Cattleya deckeri* Klotzsch. 1855.

Add under synonyms: *C. hennisiana* Charlesworth. Imported in 1888. See G. Wilson, *Amer. Orchid Soc. Bull.* 9(11):299–300, 1940.

p. 48 Note that *Cattleya deckeri* has recently been neotypified in *Lindleyana* 13(3):219, 1998, as a sketch by Reichenbach on a Decker collection. "Hort. Decker. 1853. Sheet 41961." Dressler does not mention the Curtis plate 4916, 1856, in his discussion.

p. 51 *Laeliocattleya dormaniana* (Reichenbach f.) Rolfe. 1889.

p. 55 *Cattleya virginalis* Linden and André *ex* Devos. 1876.

Cattleya wallisii Linden *ex* Reichenbach f. 1882.

Cattleya quadricolor var. *eldorado* (Linden *ex* Van Houtte) Morren and Devos. 1887.

p. 59 *Cattleya forbesii* Lindley. 1821–1824. *Coll. Bot.* 8, sub t. 37; and *Curtis's Bot. Mag.* 1833. t. 3265.

Epidendrum forbesii (Lindley) Reichenbach f. 1861.

p. 60 *Cattleya gaskelliana* (Hort. Sander *ex* N. E. Brown) Reichenbach f. *ex* B. S. Williams. 1883

Cattleya labiata var. *gaskelliana* Hort. Sander *ex* N. E. Brown. 1883.

p. 62 *Epidendrum granulosum* (Lindley) Reichenbach f. 1861.

p. 67 *Cattleya guttata* Lindley. 1831.

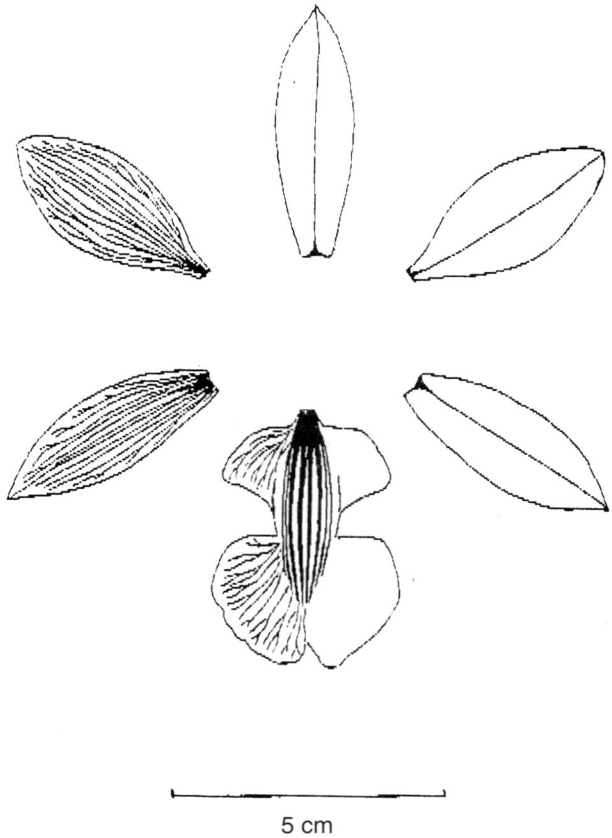

Cattleya ×joaquiniana.

Epidendrum elegans Vellozo. 1831 (1790). *Icon. Fl. Flum.* 9, t. 12.
Cattleya tigrina A. Richard *ex* Beer. 1848.
Epidendrum elatius (Lindley) Reichenbach f. 1862.

p. 72 Replace *Cattleya harrisoniae* Bateman with *Cattleya harrisoniae* Paxton. 1838. *Paxton's Mag. Bot.* 4:247.
Cattleya harrisonie P. N. Don.
Cattleya loddigesii var. *harrisoniana* Veitch.
Add under synonyms: *Cattleya harrisonii* Beer. 1854. *Prakt. Stud. an der Fam. Orchidaceae,* p. 211.
Add under synonyms: *Cattleya harrisonii* Hort. *ex* Ducharte. 1857. *Jour. Soc. d'Hort. de France 3:725, t. 7.*
Add under synonyms: *Epidendrum harrisonianum* (Bateman *ex* Lindley) Reichenbach f. 1861.
Add under synonyms: *Cattleya brownii* Rolfe. 1894. *Bull. Misc. Inform. Kew.,* p. 156.

p. 73 *Epidendrum intermedium* (Graham) Reichenbach f. 1861.
Cattleya amabilis Petot *ex* B. S. Williams. 1894. *Orchid Growers' Manual,* ed. 7, p. 152.

p. 86 *Epidendrum loddigesii* (Lindley) Reichenbach f. 1861.

p. 91 *Cattleya meyeri* Regel. 1855. *Ind. Sem. Hort. Petróp.*, p. 17.
p. 92 *Epidendrum maximum* (Lindley) Reichenbach f. 1861.
Cattleya malouana L. Linden *ex* Linden and Rodigas. 1886.
p. 94 *Cattleya cupidon* Hort. *ex* Linden and Rodigas. 1894.
p. 97 *Cattleya reineckiana* Hort. *ex* Reichenbach f. 1883.
Add under synonyms: *Cattleya reineckiana* var. *superbissima* Hast. 1889. *Gard. Chron*, n.s., 21:609.
p. 101 *Cattleya percivaliana* (Reichenbach f.) O'Brien. 1883.
p. 103 *Cattleya dijanceana* Hort. *ex* Rolfe. 1899. *Orch. Rev.* 7:332.
Add under synonyms: *Cattleya granulosa* var. *dijanceana* Veitch. 1891. *Gard. Chron.*, ser. 3, 10:310.
p. 108 *Epidendrum schillerianum* (Reichenbach f.) Reichenbach f. 1861.
p. 118 *Cattleya bogotensis* Linden *ex* Morren. 1863. *Belg. Hort.* 15:102.
p. 123 *Epidendrum superbum* (Schomburgk *ex* Lindley) Reichenbach f. 1862.
Add under synonyms: *Cattleya violacea* f. *alba* (Rolfe) Christenson. 1996. *Lindleyana* 11(1):14.
p. 125 *Epidendrum walkerianum* (Gardner) Reichenbach f. 1862.
p. 126 *Cattleya* ×*joaquiniana* Miranda is a newly described natural hybrid between *C. walkeriana* and *C. bicolor*. Please see details at p. 46 above.
p. 127 *Cattleya warneri* Moore *ex* Warner. 1862.
Consider that *Cattleya silvana* Pabst, 1976, is a separate species and should not now be listed as a doubtful or dubious species. Also see p. 134.
p 130 Change feast day of St. John to June 24.

VOLUME II. THE LAELIAS

p. 12 Add the following intergeneric combinations with *Laelia* to Table I, bringing the table up-to-date through November, 1999.

× *Barkeria* × *Brassavola* × *Cattleya* × *Epidendrum* × *Sophronitis* = *Trisuloara (Tsla.)*
× *Barkeria* × *Cattleya* × *Sophronitis* = *Matsudaara (Msda.)*
× *Brassavola* × *Broughtonia* × *Cattleya* × *Schomburgkia* = *Westara (Wsta.)*
× *Brassavola* × *Broughtonia* × *Schomburgkia* × *Sophronitis* = *Mooreara (Mora.)*
× *Brassavola* × *Diacrium* = *Gerberara (Gba.)*
× *Brassavola* × *Epidendrum* × *Laelia* × *Sophronitis* = *Panczakara (Pzka.)*
× *Broughtonia* × *Cattleya* × *Diacrium* = *Cookara (Cook.)*
× *Diacrium* × *Epidendrum* = *Gumara (Gum.)*

p. 30 *Cattleya albida* (Lindley *ex* Bateman) Beer. 1854.
Bletia albida (Lindley *ex* Bateman) Beer. 1861.
p. 32 The Mexican orchidologists now consider *Laelia eyermaniana* Reichenbach f. *pro hyb.* to be a separate species of hybrid origin. See *Laelias of Mexico* by F. Halbinger and M. Soto, 1997. Herbario AMO. México City.

p. 32 *Bletia anceps* (Lindley) Reichenbach f. 1862. *Walp. Ann.* 6:418. Add under synonyms:

Laelia anceps subsp. *anceps.* See above reference to F. Halbinger and M. Soto.

Laelia anceps subsp. *dawsonii* (J. Anderson) Rolfe. 1922. *Orch. Rev.* 30:10. Basionym: *L. anceps var. dawsonii* J. Anderson. 1868. *Gard. Chron.* 27. This subspecies has been known in the literature and trade as the "Guerrero" type. *L. dawsonii* (J. Anderson) Crawshay. 1902. *Gard. Chron.* 32:414.

p. 32 *Laelia alvaroana* Miranda. 1999. *Bradea* 8 (22): 123–125. The holotype specimen is Miranda #1389 at HB from Serra dos Orgãos near Casemiro de Abreu, Rio de Janeiro. Paratypes were collected at Santa Maria Madalena and Pedra das Flores at 1200 m. This newly described species belongs to subgenus *Parviflorae* section *Parviflorae.* It is rare, and the small flowers occur from December to March. A rupicolous species, it has previously been confused with *L. mixta.* The linear-lanceolate leaves are transversely rugose and somewhat reflexed. The inflorescence is laxly few-flowered, having deeply lobed lips with elongated, long-pointed lateral lobes exending to the sides of the convoluted lip midlobe. The flowers are yellow-orange with red veining. The sepals measure 24 × 6 mm; the petals 24 × 5 mm; and the lip midlobe about 16 mm long × 11 mm wide.

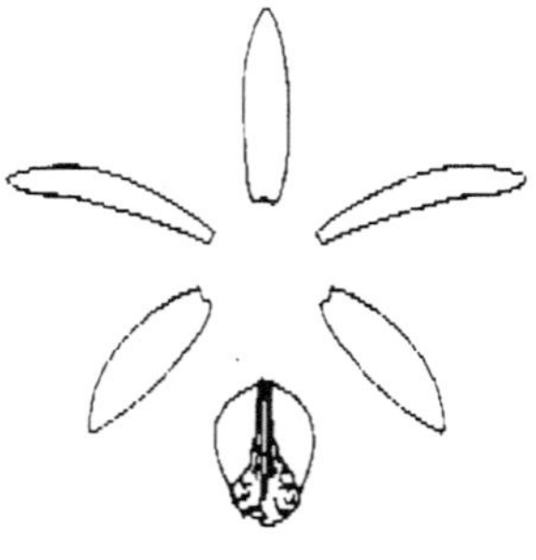

Laelia alvaroana.

p. 43 Add as a synonym: *Hoffmannseggella bahiensis* (Schlechter) Jones. 1974. *Hawaii Orch. Jour.* 3(4):15–17.

p. 50 *Laelia briegeri* Blumenschein *ex* Pabst. 1960.

p. 52 *Bletia caulescens* (Lindley) Reichenbach f. 1861.

p. 54 Add under synonyms: *Hoffmannseggella cinnabarina* (Bateman *ex* Lindley) H. G. Jones. 1968. *Acta Bot. Acad. Sci. Hung.* 14(1–2): 69.

p. 56 *Laelia crispa* (Lindley) Reichenbach f. 1828.
Cattleya reflexa Parmentier *ex* Reichenbach f. 1861.

p. 59 Add under synonyms: *Hoffmannseggella crispata* (Thunberg) H. G. Jones. 1974. *Hawaii Orch. Jour.* 3(4):15–17.

p. 60 *Laelia crispilabia* (Reichenbach f.) Warner. 1865.
Bletia crispilabia A. Richard *ex* Reichenbach f. 1863.
Add under synonyms: *Hoffmannseggella crispilabia* (A. Richard *ex* Warner) H. G. Jones. 1972. *Bradea* 1(23):266.

p. 68 *Laelia esalqueana* Blumenschein *ex* Pabst. 1960.

p. 70 *Cattleya flava* (Lindley) Beer. 1854.

p. 77 Add as a synonym: *Hoffmannseggella ghillanyi* (Pabst) H. G. Jones. 1974. *Hawaii Orch. Jour.* 3(4):16.

p. 78 Add as a synonym: *Hoffmannseggella macrobulbosa* (Pabst) H. G. Jones. 1974. *Hawaii Orch. Jour.* 3(4):15–17.

p. 82 *Laelia grandis* Lindley and Paxton. 1850.

p. 90 Add under synonyms:
Hoffmannseggella kautskyi (Pabst) H. G. Jones. 1972. *Rhodora* 74 (798):285.
Laelia kautskyi Pabst. *Orch. Rev.* 78:321–323.

p. 93 Add as synonym of *Laelia liliputana: Hoffmannseggella liliputana* (Pabst) H. G. Jones. 1974. *Hawaii Orch. Jour.* 3(4):15–17.

p. 96 *Laelia longipes* (Reichenbach f.) Cogniaux. 1863. *Xenia Orch.*
Bletia longipes (Reichenbach f.) Reichenbach f. 1863. *Xenia Orch.*
Transfer *Bletia lucasiana* to synonyms of *Laelia lucasiana,* p. 97, where it should be *Bletia lucasiana* (Rolfe) Reichenbach f. 1897. *Orchid Review* 5:112.
Add under synonyms: *Laelia longipes* var. *alba* Rolfe. 1897. *Orch. Rev.* 5:112.

p. 98 *Laelia lundii* (Reichenbach f. and Warming) Cogniaux. In *Martius* 3(5):301.

p. 101 *Laelia milleri* Blumenschein *ex* Pabst. 1960.

p. 104 *Laelia munchowiana* Miranda. 1999. *Bradea* 8 (22): 126–128. This newly described species of rupicolous *Laelia* belongs in subgenus *Parviflorae* section *Rupestres*. It is from the state of Espírito Santo in the Serrado do Crenaque near Baixo Guandu. The plants produce up to 20 flowers, rose-magenta with darker veining. It is the first rose-colored species to be found in Espírito Santo and is near its border with Minas Gerais. The dark green leaves are leathery and with purple on the borders and lower surfaces. Measurements show cylindrical pseudobulbs 4 cm tall × 1.5 cm wide; leaves 9 cm long × 2.2 cm wide; spathes 4 × 7 mm; inflorescences successively flowering 50 cm tall; linear-lanceolate sepals 12–15 mm × 3–3.5

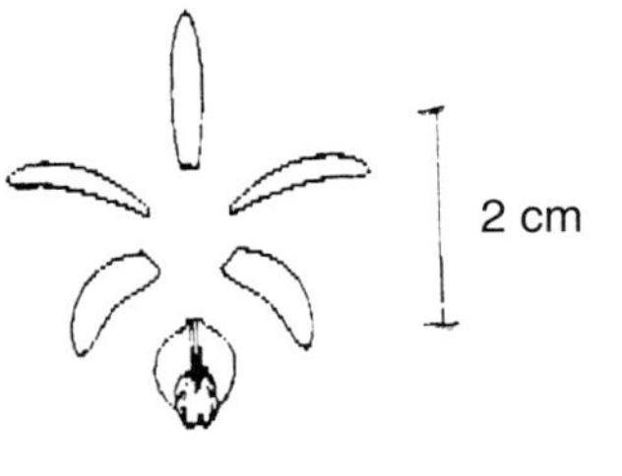

Laelia munchowiana.

mm; linear-oblanceolate petals 14 × 2.2 mm; and lip about 10 mm long and 7.5 mm wide across the midlobe.

p. 104 *Laelia perrinii* (Lindley) Bateman. 1842.

p. 107 *Cattleya pinellii* Hort. *ex* Lindley. 1844.
Bletia pumila (Hooker) Reichenbach f. 1861.

p. 110 *Cattleya purpurata* (Lindley and Paxton) Beer. 1854.
Bletia purpurata (Lindley and Paxton) Reichenbach f. 1861.

p. 119 Add under synonyms: *Laelia inconspicua* H. G. Jones. 1974. *Adansonia,* ser. 2, 14(2):299.

p. 120 The yellow forms that were tentatively designated as *Laelia rubescens* var. *aurea* are now *Laelia aurea* Navarro *ex* Halbinger and Salazar. 1990. *Orquídea* (Méx.) 12(1): 41–46. Figs.

p. 125 Add under synonyms: *Laelia grandiflora* var. *alba* Rolfe. 1906. *Orch. Rev.* 14:254.

p. 133 Add as a synonym: *Hoffmannseggella tereticaulis* (Hoehne) H. G. Jones. 1974. *Florida Orchidist* 4:156.

p. 147 Add *fournieri* (variety of *lucasiana*) 97. Also add it to Plant Name Index.

VOLUME III. *SCHOMBURGKIA, SOPHRONITIS*, AND OTHER SOUTH AMERICAN GENERA

p. 12 Add the following intergeneric combinations with *Schomburgkia* to Table I, bringing the table up-to-date through November, 1999.

× *Brassavola* × *Broughtonia* × *Cattleya* × *Laelia* = *Westara (Wsta.)*
× *Brassavola* × *Broughtonia* × *Cattleya* × *Laelia* × *Sophronitis* = *Mooreara (Mora.)*
× *Brassavola* × *Cattleya* × *Epidendrum* = *Symmonsara (Syma.)*

p. 25 Add Brazil to countries of distribution for *Schomburgkia crispa.* Recent work by Lou C. Menezes has demonstrated to me that *S. crispa,* not *S. gloriosa* nor *S. fimbriata,* accounts for most of the *Schomburgkia* specimens currently found in Brazil.

p. 29 Add under synonyms: *Schomburgkia amazonica* Hoehne *ex* H. G. Jones. 1967. *Repert. Spec. Nov. Regni Veg.* 76:197.

p. 52 Add under synonyms:
Epidendrum wendlandianum Kraenzlin. 1893. *Gard. Chron.,* ser. 3, 14:58.
Epidendrum ensicaulon Richard and Galeotti. 1862. *Walp. Ann. Bot.* 6:351.
Epidendrum venosum Lindley. 1831. *Gen. and Sp. Orch. Pl. Epidendreae,* p. 99.
Encyclia venosa Schlechter. 1918. *Beih. Bot. Centralbl.* 36(2):474.
Encyclia wendlandiana (Kraenzlin) Schlechter. 1918. *Beih. Bot. Centralbl.* 36(2):474.

p. 60 Add the following intergeneric combinations with *Sophronitis* to Table III, bringing the table up-to-date through November, 1999.

× *Barkeria* × *Brassavola* × *Cattleya* × *Epidendrum*
= *Trisuloara (Tsla.)*
× *Barkeria* × *Cattleya* × *Laelia* = *Matsudaara (Msda.)*
× *Brassavola* × *Broughtonia* × *Cattleya* × *Laelia* × *Schomburgkia*
= *Mooreara (Mora.)*
× *Brassavola* × *Cattleya* × *Diacrium* = *Clarkeara (Clka.)*
× *Brassavola* × *Epidendrum* × *Laelia* = *Panczakara (Pzka.)*

p. 74 *Sophronitis mantiqueirae* (Fowlie) Fowlie. 1972.

p. 76 *Sophronitis pterocarpa* Lindley and Paxton. 1853.

p. 88 Cogniaux did not "miss" the genus *Isabelia* in his treatment of Brazilian orchids. It is covered in his Addendum at the end of that critical and valuable work.

p. 95 Add the following paragraph to the introductory section of *Leptotes:*

Leptotes (Lpt.), in particular *Lpt. bicolor,* has been used to produce a few intergeneric hybrids. Most are bigeneric, and only one to date is trigeneric. They are *Leptokeria (Lptka.)* with *Barkeria, Leptovola (Lptv.)* with *Brassavola, Epileptovola (Elva.)* with *Epidendrum* and *Brassavola, Cattotes (Ctts.)* with *Cattleya, Leptodendrum (Lptdm.)* with *Epidendrum,* and *Leptolaelia (Lptl.)* with *Laelia.*

p. 95 Add as a synonym: *Bletia bicolor* (Lindley) Reichenbach f. 1864. *Walp. Ann.* 6:438.

p. 95 Eric Christenson advises me that the species *Leptotes mogyensis* by F. J. Krackowizer in *Revista do Círc. Paul. de Orquidóf.* 11(4): 64, 1954, is also a good species but invalidly published as there is no type specimen cited nor conserved in a recognized herbarium.

p. 96 Add as synonyms of *Leptotes serrulata* Lindley:

Leptotes bicolor var. *serrulata* (Lindley) Reichenbach f. 1857. *Cat. Orch. Schill.,* p. 40; also B. Stein, 1892, *Orchideenbuch,* p. 307.

Tetramicra serrulata (Lindley) Bentham and J. D. Hooker. In D. Bois, 1893, *Les Orchidées Manuel de l'Amateur,* p. 93.

p. 97 Eric Christenson advises me that *Leptotes minuta* Rolfe, listed as a synonym under *Lpt. tenuis,* is a separate species and should be considered as such. The type description is as listed in the *Gardeners' Chronicle* in 1889.

p. 101 *Pinelia* should be changed to *Pinelianthe* because of the similarity between the two genera *Pinelia* and *Pinellia.*

p. 105 See Volume V, p. 152, for a new treatment of Lehmann's *Pinelia.* Add as synonyms:

Restrepia lehmanniana (Kraenzlin) Schlechter. 1907. *Repert. Spec. Nov. Regni Veg.* 3:277.

Pinelia lehmanniana Kraenzlin. 1899. *Bot. Jahrb. Syst.* 26(5): 479.

p. 115 Check Dressler in *Novon* 7(2):124, 1997, for new complications with the former concept of *Reichenbachanthus lankesteri.*

VOLUME IV. THE BAHAMIAN AND CARIBBEAN SPECIES

p. 17 Add the following intergeneric combinations with *Broughtonia* to Table I, bringing the table up-to-date through November, 1999.

× *Brassavola* × *Cattleya* = *Carmichaelara* (*Crml.*)
× *Brassavola* × *Cattleya* × *Laelia* × *Schomburgkia* = *Westara* (*Wsta.*)
× *Brassavola* × *Laelia* × *Schomburgkia* × *Sophronitis* = *Mooreara* (*Mora.*)
× *Epidendrum* × *Laeliopsis* × *Tetramicra* = *Balaguerara* (*Blga.*)

p. 44 Add the following intergeneric combinations with *Epidendrum* (*Encyclia*) to Table IV, bringing it up-to-date through November, 1999.

× *Barkeria* × *Brassavola* × *Cattleya* × *Laelia* × *Sophronitis* = *Trisuloara* (*Tsla.*)
× *Brassavola* × *Cattleya* × *Schomburgkia* = *Symmonsara* (*Syma.*)
× *Brassavola* × *Laelia* × *Sophronitis* = *Panczakara* (*Pzka.*)
× *Brassavola* × *Leptotes* = *Epileptovola* (*Elva.*)
× *Broughtonia* × *Laeliopsis* × *Tetramicra* = *Balaguerara* (*Blga.*)
× *Diacrium* × *Laelia* = *Gumara* (*Gum.*)

p. 51 *Encyclia bradfordii* (Grisebach) Carnevali and Ramírez. 1986. *Ernestia* 36:9.

p. 83 Add the following intergeneric combination with *Laeliopsis* to Table VI, bringing the table up-to-date through November, 1999.

× *Broughtonia* × *Epidendrum* × *Tetramicra* = *Balaguerara* (*Blga.*)

p. 117 Add the following intergeneric combination with *Tetramicra* to Table VII, bringing the table up-to-date through November, 1999.

× *Broughtonia* × *Epidendrum* × *Laeliopsis* = *Balaguerara* (*Blga.*)

p. 120 Add under synonyms: *Brassavola rigida* I. Boldingh. 1909. *The Flora of the West Dutch Indian Islands*, 1:43.

Plate 44 should be *Psychilis macconnelliae* from the Virgin Islands
Plate 45 should be *Psychilis atropurpurea* from the Dominican Republic.
Plate 46 credit should be Carl L. Withner, not Ruben Sauleda.
Plate 54 is the unusual pink form of *P. macconnelliae* from Vieques Island.

VOLUME V. *BRASSAVOLA*, *ENCYCLIA*, AND OTHER GENERA OF MÉXICO AND CENTRAL AMERICA

Note: *Dimeranda* should be *Dimerandra* throughout.

p. 21 Add the following intergeneric combination with *Barkeria* to Table I, bringing the table up-to-date through November, 1999.

× *Brassavola* × *Cattleya* × *Epidendrum* × *Laelia* × *Sophronitis* = *Trisuloara* (*Tsla.*)

p. 36 Add the following intergeneric combinations with *Brassavola* to Table II, bringing the table up-to-date through November, 1999.

× *Barkeria* × *Cattleya* × *Epidendrum* × *Laelia* × *Sophronitis* = *Trisuloara (Tsla.)*
× *Broughtonia* × *Cattleya* × *Laelia* × *Schomburgkia* = *Westara (Wsta.)*
× *Broughtonia* × *Laelia* × *Schomburgkia* × *Sophronitis* = *Mooreara (Mora.)*
× *Cattleya* × *Broughtonia* = *Carmichaelara (Crml.)*
× *Cattleya* × *Diacrium* × *Sophronitis* = *Clarkeara (Clka.)*
× *Cattleya* × *Epidendrum* × *Schomburgkia* = *Symmonsara (Syma.)*
× *Epidendrum* × *Laelia* × *Sophronitis* = *Panczakara (Pzka.)*
× *Epidendrum* × *Leptotes* = *Epileptovola (Elva.)*

p. 40 Add to "Index to Species of *Brassavola*" the following:
Brassavola ovaliformis, 48

p. 46 *Brassavola ovaliformis* Schweinfurth, 1949, should be removed from synonymy with *Brassavola chacoensis* and considered a species in its own right. Research with fresh flowers provided by Dr. Isaias Rolando of Peru has enabled me to make the distinction. Close examination of the fresh column apex reveals a long tooth or ligule (visible in the drawing of the column tip in Figure 5-4). This feature seems unique for this species and to date could be considered a hallmark feature for the species. It is not mentioned in the type description of the species. The flowers are not fragrant during the day, but they do have a weak , somewhat cinnamony fragrance at night.

p. 48 Caption for Figure 5-4 should now read "*Brassavola ovaliformis.* Drawing from Schweinfurth's *Orchids of Peru.* Courtesy of *Fieldiana, Botany*, Field Museum of Natural History. Chicago, Illinois."

p. 58 Beware that *Brassavola rhomboglossa* may be a separate species and not a synonym under *Brassavola perrinii.* The drawing of the type specimen by Pabst in *Bradea* shows pubescence of the lower parts of the sepals and also the lower surface of the column. The anther also has a unique shape and attachment to the column with apparently a long tooth or ligule at the apex. Until fresh flowers are available for study, these details cannot be checked with accuracy. Pabst suggested that the flower is a closest to *Brassavola cebolleta*, a confusing suggestion in my opinion. It is even more likely to be *B. ovaliformis* (see p. 46 above).

p. 67 Add Saba (West Indies) to distribution of *Brassavola venosa.*

p. 71 Add the following intergeneric combination with *Diacrium* (*Caularthron)* to Table III, bringing the table up-to-date through November, 1999.

× *Brassavola* × *Cattleya* × *Sophronitis* = *Clarkeara (Clka.)*
× *Epidendrum* × *Laelia* = *Gumara (Gum.)*

p. 80 Center figure is *D. latipetala*, not *D. lamellatum.* Lower left figure is *D. stenopetala*, not *D. stenopetalum.*

p. 122 *Encyclia peraltensis* (Ames) Dressler. 1997. *Novon* 7(2):124.

p. 123 The synonym *Encyclia* ×*perplexa* should be *Enc. perplexa* (Dressler and Pollard) Withner. 1998. *Stat. nov.* The situation here is that *perplexa* was changed from a species to a natural hybrid of two species and transferred to the genus *Encyclia* by Dressler and Pollard. The basionym is *Epidendrum perplexa* Ames, Hubbard, and Schweinfurth. I now want to change it from a hybrid to a species. To complicate matters, Hoehne transferred *Epidendrum oncidioides* var. *perplexa* to the genus *Encyclia* in 1952, predating Dressler and Pollard's use of *Encyclia* in 1971. Dressler and Pollard did not mention Hoehne. Furthermore, in 1971 they wrote *comb. & stat nov.*, but should have written only *stat. nov.*

p. 138 *Euchile citrina* (Dressler) Withner.

p. 186 *Brassavola chacoensis*, 40–42, 46–47, 60

p. 186 *Brassavola ovaliformis*, 39, 40, 46, 48

Plate 28 should be *Brassavola ovaliformis.*

Plate 88 should be *Homalopetalum pachyphyllum.*

VOLUME VI. THE SOUTH AMERICAN *ENCYCLIA* SPECIES

p. 4 Add to end of first full paragraph: Long ago Stalky Dunsterville sent me photocopies of several *Encyclia* plates he had drawn. I am using them in this volume to illustrate some of the species I now believe they portray, and finally give those plates the benefit of a more current opinion on their identification.

Selected References for Additional Reading

Adams, H. and E. Anderson. 1958. A conspectus of hybridization in the Orchidaceae. *Evolution* 12:512.

Ames, O., T. F. Hubbard, and C. Schweinfurth. 1935. A fourth polymorphic alliance in *Epidendrum. Bot. Mus. Leaflt., Harv. Univ.* 3:93–110.

Ames, O., T. F. Hubbard, and C. Schweinfurth. 1936. *The Genus* Epidendrum *in the United States and Middle America.* Botanical Museum, Harvard University. Cambridge, Massachusetts.

Ben-Avraham, Zvi. 1981. The movement of continents. *American Scientist* 69 (May–June):291–299.

Bennett, D. E. and E. Christenson. 1995. *Icones Orchidacearum Peruviarum.* A. Pastorelli de Bennett, Lima. Peru.

Benzing, D. 1987. Major patterns and processes in orchid evolution: a critical synthesis. *Orchid Biology, Reviews and Perspectives* 4:33–77.

Brieger, F. G. 1960a. Contribuiçóes para a taxonomia das orquideas. *Publicacão Cientifica, Instituto de Genetica, Universidade de São Paulo* 1:1.

Brieger, F. G. 1960b. Geographic distribution and phylogeny of orchids. *Proc. 3rd World Orchid Conf.*, p. 328.

Brieger, F. G. 1961. On the taxonomy of the tribe Laelieae (Orchidaceae). *Proc. 3rd World Orchid Conf.*, p. 333.

Brieger, F. G. 1973. Sobre a distribuicão fitogeografica das orquideas brasileiras. *Orquídea* (Brazil) 30(2):76.

Burns-Balogh, P. and V. Funk. 1986. A phylogenetic analysis of the Orchidaceae. *Smithsonian Contrib. to Bot.* no. 61.

Camp, W. H. 1947. Distribution patterns in modern plants and the problems of ancient dispersals. *Ecol. Monographs* 17:123, 159.

C.A.O.B. = Coordenadoria Associaçóes Orquidófilas do Brasil.

Carnevali, G. and I. Ramírez. 1986. *Encyclia sclerocladia. Ernestia* 36:9.

Carnevali, G. and I. Ramírez. 1988. Revisión del genero *Encyclia* para Venezuela. *XLVI Exposicíon de Orquídeas, Comité de Orquideología,* Bol. No. 23:13–87.

Carnevali, G., I. R. de Carnevali, and G. Romero. 1994. *Orchidaceae Dunstervillorum* VIII. New *Encyclia* species and combinations from Venezuelan Guayana. *Lindleyana* 9(1):59–70.

Castro Neto, Victorino Paiva. 1998. *Encyclias brasileiras.* Part I. *Orquidário* 12(2):44.

Chase, M. 1986. A reappraisal of the oncidioid orchids. *Syst. Bot.* 11:477.

Chase, M. and H. Hills. 1992. Orchid phylogeny, flower sexuality, and fragrance-seeking. *Bioscience* 42:43.

Cogniaux, A. 1893–1906. Orchidaceae II, Tribus VII Laeliinae. In Martius's *Flora Brasiliensis* 3(5) (*Martius* in text).

Cracraft, J. 1983. Cladistic analysis and vicariance biogeography. *American Scientist* 71:273.

Day, J. 1863–1888. 53 oblong books, about 90 pages each, bound in 17 volumes, indexed. They contain about 3000 colored drawings, 500 in sepia, original notes, and clippings from the *Gardeners' Chronicle* and other sources. Consult W. B. Hooker in *Bull. Misc. Info.*, Royal Bot. Gardens, Kew, 1906, for other details about John Day.

Dodson, C. H. 1988. A list of the orchid species reported for Ecuador. In A. Arosemena et al., *Orquídeas de la Costa del Ecuador.* Asociación Ecuatoriana de Orquideología. Guayaquil, Ecuador.

Dodson, C. H. 1996. To the rescue of paradise. In M. Ospina H., *Orchids and Ecology in Colombia.*

Dodson, C. H. and P. Dodson. 1980. *Icones Plantarum Tropicarum. Orchids of Ecuador.* The Marie Selby Botanical Gardens. Sarasota, Florida.

Dressler, R. L. 1962. Is it really *Epidendrum oncidioides? Florida Orchidist* 3:124.

Dressler, R. L. 1981. *The Orchids, Natural History and Classification.* Harvard University Press, Cambridge, Massachusetts.

Dressler, R. L. 1993. *Phylogeny and Classification of the Orchid Family.* Dioscorides Press. Portland, Oregon.

Dressler, R. L. and G. E. Pollard. 1974. *The Genus* Encyclia *in México.* Asociación Mexicana de Orquideología. México 17, D. F., México.

Dunsterville, G. C. K. and L. Garay. 1959–1976. *Venezuelan Orchids Illustrated.* 6 vols. Andre Deutsch. London.

Dunsterville, G. C. K. and L. Garay. 1979. *Orchids of Venezuela: An Illustrated Field Guide.* 3 vols. Botanical Museum, Harvard University. Cambridge, Massachusetts.

Escobar, Rodrigo R. 1991. *Native Colombian Orchids* 2:166.

Escobar, Rodrigo R. 1994. *Native Colombian Orchids* 5:751.

Foldats, E. 1970. Orchidaceae. In T. Lasser, *Flora of Venezuela* 15(3).

Fowlie, J. A. 1992. In Brazil. Part 38. A habitat of *Encyclia dichroma* on rocky granite outcrops in interior of Bahia, Brazil. *Orchid Digest* 56(3):139–141.

Fowlie, J. A. 1993. In Brazil. Part 39. *Encyclia ghillanyi* in a caatinga in Sergipe. *Orchid Digest* 57(1):13–15.

Fowlie, J. A. and D. Duveen. 1992. A contribution to an understanding of the genus *Encyclia* as it occurs in the Brazilian Shield and its river tributaries. *Orchid Digest* 56(4):171–206.

Garay, L. 1960. On the origin of the Orchidaceae. *Proc. 3rd World Orchid Conf.*, p. 172.

Garay, L. 1963. Evolutionary significance of geographical distribution of orchids. *Proc. 4th World Orchid Conf.*, p. 170.

Haffer, J. 1969. Speciation in Amazonian forest birds. *Science* 165(3889):131.

Hamilton, R. M. 1986. *When Does It Flower?* 2nd ed. Published by the author. Vancouver, British Columbia, Canada.

Higgins, W. E. 1997. A reconsideration of the genus *Prosthechea* (Orchidaceae). *Phytologia* 82 (5): 370–383.

Hoehne, F. C. 1949. *Iconografia de Orchidáceas do Brasil.* Secretaria de Agriculturia. São Paulo, Brazil.

Hoehne, F. C. 1952. Lista preliminar das especies do genero *Encyclia* Hooker para controle. *Arq. Bot. Est. São Paulo,* n.s., 2(6):151.

Holttum, R. 1955. Growth habits of monocotyledons—variations on a theme. *Phytomorphology* 5:399.

Hurley, Patrick M. 1968. The confirmation of continental drift. *Scientific American* (April):53–64.

Kirschvink and Ripperdam. 1997. *Science* (July 25).

Lasser, Thomas. 1970. *Flora de Venezuela* 15:85.

Martius. See Cogniaux 1893–1906.

Mansfeld, R. 1929. *Figuren-Atlas zu den Orchideenenfloren der südamerikanischen Kordillerenstaaten von R. Schlechter. Repert. Spec. Nov. Regni Veg.,* Beihefte LVII:VI–X.

Mansfeld, R. 1930. *Blütenanalysen neuer Orchideen von R. Schlechter. Repert. Spec. Nov. Regni Veg.,* Beihefte LVIII:I–IV.

Menezes, L. C. 1992. Variation in *Encyclia lutzenbergeri* L. C. Men. and its affinity with *Encyclia osmantha* (B. Rodrigues) Porto and Brade. *Orchid Digest* 56(1):20.

Merritt, J. I. 1979. Hess's geological revolution. *Princeton Alumni Weekly* 80(2):10.

Mooney, W. 1981. An East Pacific–Caribbean Ridge during the Jurassic and Cretaceous and the evolution of western Colombia. In M. McElhinny and D. Valencio, eds., *Paleoreconstruction of the Continents.* Geodynamics series, vol. 2:55. American Geophysical Union. Washington, D.C.

Moore, T. 1857. *Illustrations of Orchidaceous Plants.* Willas and Sotheran. London.

Ortíz, P. 1995. *Orquídeas de Colombia.* 2nd ed. Corporación Capitalina de Orquideología. Santefé de Bogotá, Colombia.

Ospina H., M. 1996. *Orchids and Ecology in Colombia.*

Ossian, Clair. 1994. Origin of the Orchidaceae–the geological evidence. *Proc. 14th World Orchid Conf.*, p. 105.

Pfitzer, E. 1889. Orchidaceae. In A. Engler and K. Prantl, *Die Naturliche Pflanzenfamilien. Orchidaceae* 2(6):140–149.

Porto, P. C. and A. C. Brade. 1935. Index Orchidacearum in Brasilia. *Rodriguésia* 1(2):28–32.

Prance, G. 1973. Phytogeographic support for the theory of Pleistocene forest refuges in the Amazon Basin, based on distribution patterns. *Acta Amazonica* 3(3):5.

Prance, G. 1985. The changing forests. In G. Prance and E. Lovejoy, eds., *Amazonica,* p. 146. Pergamon Press.

Pridgeon, A. M. and M. Chase. 1998. Phylogenetics of subtribe Catasetinae

(Orchidaceae) from nuclear and chloroplast DNA sequences. *Proc. 15th World Orchid Conf.*, p. 275.

Rafinesque, C. S. 1836 (1837). *Exophya. Flora Telluriana* 2:63.

Rafinesque, C. S. 1836 (1838). *Doxosma, Psychilis. Flora Telluriana* 4:9.

Ragan, M. 1998. Encyclias of the Caribbean and the Bahamas. In *Proc. 15th World Orchid Conf., Rio de Janeiro 1996*, pp. 245–247. Naturalia Publications, Transfaire S. A. Turriers. France.

Raven, P. and D. Axelrod. 1974. Angiosperm biogeography and past continental movements. *Ann. Mo. Bot. Gard.* 61(3):539.

Raven, P. and D. Axelrod. 1975. History of the flora and fauna of Latin America. *American Scientist* 63:420.

Rodrigues da Silveira, F. 1935. Genera Orchidacearum Brasiliensis. *Rodriguésia* 1(2):13.

Rodríguez, R. L. with D. E. Mora, M. E. Barahona, and N. H. Williams. 1986. *Géneros de Orquídeas de Costa Rica.* Editorial Universidad de Costa Rica. San José, Costa Rica.

Romero, G. 1993a. Unique orchid habitats in southern Venezuela. *Am. Orchid Soc. Bull.* 62(7):698.

Romero, G. 1993b. Unique orchid habitats in southern Venezuela. II. Arbustales or white-sand shrublands. *Am. Orchid Soc. Bull.* 62(8):811.

Scotese, C. R. 1990. *Atlas of Phanerozoic Plate Tectonic Reconstructions.* International Lithosphere Program. Paleomap Project Technical Report no. 10-90-1. Department of Geology, University of Texas at Arlington.

Smith, L. B. 1962. Origins of the flora of southern Brazil. *Contrib. U.S. Natl. Herbarium* 35(3):215.

Sprunger, S., P. Cribb, and A. Toscano de Brito, eds. 1996. João Barbosa Rodrigues. *Iconographie des Orchidées du Brésil.* Volume I: The Illustrations. Volume II: The Texts. Friedrich Reinhardt Verlag. Basel, Switzerland.

Veitch and Sons, J. 1887. *Manual of Orchidaceous Plants. Epidendrum*, p. 86. Chelsea.

Werkhoven, M. C. M. 1986. *Orchids of Suriname.* VACO N. V. Paramaribo, Suriname.

Williams, L. O. 1939. Las Orchidáceas del Noroeste Argentino. *Lilloa* 4(2): 337.

Withner, C., ed. 1974. *The Orchids, Scientific Studies.* John Wiley and Sons. New York.

Withner, C. 1980. *Encyclia viridiflora*—The First Encyclia (Or How It All Began). *Amer. Orchid Soc. Bull.* 49(7):756.

Withner, C. and H. Adams. 1960. Generic relationships and evolution among the cattleyas and their relatives. *Proc. 3rd World Orchid Conf.*, p. 136.

Plant Name Index

Index of Persons

General Index